RECONCILIATION

BENAZIR BHUTTO

RECONCILIATION

Islam, Democracy, and the West

HARPER

An Imprint of HarperCollins*Publishers*
www.harpercollins.com

HarperCollins books may be purchased for educational, business, or sales promotional use. For information, please write: Special Markets Department, HarperCollins Publishers, 10 East 53rd Street, New York, NY 10022.

FIRST EDITION

Designed by Leah Carlson-Stanisic

Library of Congress Cataloging-in-Publication Data is available upon request.

ISBN: 978-0-06-156758-2

08   09   10   11   12   DIX/RRD   10   9   8   7   6   5   4   3   2   1

Contents

Note to the Reader

This book was written under extraordinary circumstances. It was my privilege to work with Benazir Bhutto on this project over the last very difficult months. This period of her life included her historic return to Karachi on October 18, 2007, which attracted three million supporters to greet her, and the unsuccessful assassination attempt on her in the early minutes of October 19 that killed 179 people. In the midst of all of this tumult Benazir and I collaborated on the book, at times while Benazir was under house arrest by the Musharraf regime and under the constraints of emergency rule, tantamount to martial law.

Despite the events swirling about her and her responsibility of leading Pakistan's largest political party—the Pakistan Peoples Party—in the parliamentary election campaign, Benazir Bhutto remained focused on *Reconciliation: Islam, Democracy, and the West*. This book was very important to her, and she threw herself into it with the complete attention and intensity with which she did so many things in her life. Benazir was convinced that the battles between democracy and dictatorship, and between extremism and moderation, were the two central forces of the new millennium. She believed that the message of her cherished religion, Islam, was being politicized and exploited by extremists and fanatics. And she believed that under dictatorship, extremism festered and grew, threatening not only her homeland of Pakistan but also the entire world.

That's why she wrote this book. That's why it was so important to her. And that's why she devoted herself totally to this project, quite literally until the early morning of her death, when I received her final edits of the manuscript.

Although I helped Benazir research and write this book, it is her work from beginning to end: a positive statement of reconciliation among religions and nations; a bold assertion of the true nature of Islam; and a practical road map for bringing societies together.

Benazir Bhutto was the bravest person I have ever known and a dear, irreplaceable friend. She was assassinated on December 27, 2007, in Rawalpindi, Pakistan. I find some solace in knowing that the last memory of her will not be the bloody carnage of the murder scene, but rather the legacy of this book, which manifests the strength, optimism, and vision of a great woman.

MARK A. SIEGEL
Washington, D.C.
December 28, 2007

1

The Path Back

As I stepped down onto the tarmac at Quaid-e-Azam International Airport in Karachi on October 18, 2007, I was overcome with emotion. Like most women in politics, I am especially sensitive to maintaining my composure, to never showing my feelings. A display of emotion by a woman in politics or government can be misconstrued as a manifestation of weakness, reinforcing stereotypes and caricatures. But as my foot touched the ground of my beloved Pakistan for the first time after eight lonely and difficult years of exile, I could not stop the tears from pouring from my eyes and I lifted my hands in reverence, in thanks, and in prayer. I stood on the soil of Pakistan in awe. I felt that a huge burden, a terrible weight, had been lifted from my shoulders. It was a sense of liberation. I was home at long last. I knew why. I knew what I had to do.

I had departed three hours earlier from my home in exile, Dubai. My husband, Asif, was to stay behind in Dubai with our two daughters, Bakhtawar and Aseefa. Asif and I had made a very calculated, difficult decision. We understood the dangers and the risks of my return, and we wanted to make sure that no matter what happened, our daughters and our son, Bilawal (at college at Oxford), would have a parent to take care of them. It was a discussion that few husbands and wives ever have to have, thankfully. But Asif and I had become accustomed to a life of sacrificing our personal happiness and any sense of normalcy and privacy. Long ago I had made my choice. The people of Pakistan have always come first. The people of Pakistan will always come first. My children understood it and not only accepted it but encouraged me. As we said good-bye, I turned to the group of assembled supporters and press and said what was in my heart: "This is the beginning of a long journey for Pakistan back to democracy, and I hope my going back is a catalyst for change. We must believe that miracles can happen."

The stakes could not have been higher. Pakistan under military dictatorship had become the epicenter of an international terrorist movement that had two primary aims. First, the extremists' aim to reconstitute the concept of the caliphate, a political state encompassing the great Ummah (Muslim community) populations of the world, uniting the Middle East, the Persian Gulf states, South Asia, Central Asia, East Asia, and parts of Africa. And second, the militants' aim to provoke a clash of civilizations between the West and an interpretation of Islam that rejects pluralism and modernity. The goal—the great hope of the militants—is a collision, an explosion between the values of the West and what the extremists claim to be the values of Islam.

Within the Muslim world there has been and continues to be an internal rift, an often violent confrontation among sects, ideologies, and interpretations of the message of Islam. This destructive tension has set brother against brother, a deadly fratricide that has tortured intra-Islamic relations for 1,300 years. This sectarian conflict stifled the brilliance of the Muslim renaissance that took place during the Dark Ages of Europe, when the great universities, scientists, doctors, and artists were all Muslim. Today that intra-Muslim sectarian violence is most visibly manifest in a senseless, self-defeating sectarian civil war that is tearing modern Iraq apart at its fragile seams and exercising its brutality in other parts of the world, especially in parts of Pakistan.

And as the Muslim world—where sectarianism is rampant—simmers internally, extremists have manipulated Islamic dogma to justify and rationalize a so-called jihad against the West. The attacks on September 11, 2001, heralded the vanguard of the caliphate-inspired dream of bloody confrontation; the Crusades in reverse. And as images of the twin towers burning and then imploding were on every television set in the world, the attack was received in two disparate ways in the Muslim world. Much, if not most, of the Muslim world reacted with horror, embarrassment, and shame when it became clear that this greatest terrorist attack in history had been carried out by Muslims in the name of Allah and jihad. Yet there was also another reaction, a troubling and disquieting one: Some people danced in the streets of Palestine. Sweets were exchanged by others in Pakistan and Bangladesh. Condemnations were few in the world's largest Muslim nation, Indonesia. The hijackers of September 11 seemed to touch a nerve of Muslim impotence. The burning and then collapsing towers represented, to some, resurgent Muslim power, a perverse Muslim payback for the domination of the West. To others it was a religious epiphany. And to still others it combined political, cultural, and religious assertiveness. A Pew comparative study of Muslims' attitudes after the attacks found that people in many Muslim countries "think it is good that Americans now know what it is like to be vulnerable."

One billion Muslims around the world seemed united in their outrage at the war in Iraq, damning the deaths of Muslims caused by U.S. military intervention without U.N. approval. But there has been little if any similar outrage against the sectarian civil war, which has led to far more casualties. Obviously (and embarrassingly), Muslim leaders, masses, and even intellectuals are quite comfortable criticizing outsiders for the harm inflicted on fellow Muslims, but there is deadly silence when they are confronted with Muslim-on-Muslim violence. That kind of criticism is not so politically convenient and certainly not politically correct. Even regarding Darfur, where there is an actual genocide being committed against a Muslim population, there has been a remarkable absence of protests, few objections, and no massive coverage on Arab or South Asian television.

We are all familiar with the data that pour forth from Western survey research centers and show an increasing contempt for and hostility to the West, and particularly the United States, in Muslim communities from Turkey to Pakistan. The war in Iraq is cited as a reason. The situation in Palestine is given as another reason. So-called decadent Western values are often part of the explanation. It is so much easier to blame others for our problems than to accept responsibility ourselves.

The colonial experience has obviously had a major impact on the Muslim psyche. Colonialism, resource exploitation, and political suppression have affected Muslims' attitudes toward the West and toward themselves. No one doubts that the record of the West in majority Islamic nations is not a pretty one. But what outsiders did in the past does not exclusively account for the quality of Muslim life today. There are a rush and an ease to condemn foreigners and colonizers, but there is an equally weighty unwillingness within the Muslim world to look inward and to identify where we may be going wrong ourselves.

It is uncomfortable but nevertheless essential to true intellectual dialogue to point out that national pride in the Muslim world is rarely derived from economic productivity, technical innovation, or intellectual creativity. Those factors seem to have been part of the Persian, Mughal, and Ottoman past but not the Muslim present. Now we see Muslim pride always characterized in the negative, derived from notions of "destroying the enemy" and "making the nation invulnerable to Western assault." Such toxic rhetoric sets the stage for the clash of civilizations between Islam and the West every bit as much as do Western military or political policies. It also serves as an opiate that keeps Muslims angry against external enemies and allows them to pay little attention to the internal causes of intellectual and economic decline. Reality and intellectual honesty demand that we look at both sides of the coin.

The burning twin towers have become a dual metaphor for both the intra-Islamic debate about the political and social values of democracy and modernity and the looming potential for a catastrophic showdown between Islam and the West. And for both of these epic battles, my homeland of Pakistan has become the epicenter—the ground zero if you will—of either reconciliation or disaster.

Few on the airplane that carried me from Dubai to Karachi on that fateful day in October 2007 knew that in my briefcase I carried with me the manuscript of this book exploring the dual crises confronting the Islamic world—both internal and external. Within hours of my reaching Pakistan, some of the pages of this book would be symbolically charred by fire and splattered with the blood and flesh of disembodied innocents thrown up by devastating terrorist bombs. The carnage that accompanied the joyous celebration of my return was a horrific metaphor for the crisis that lies before us and the need for an enlightened renaissance both within Islam and between Islam and the rest of the world.

When I had returned to Pakistan in 1986 after two years abroad, I was greeted by enormous crowds swelling to one million in Lahore, the capital city of Pakistan's most populace province, the Punjab. The size of those crowds was interpreted as an indicator of the support of the Pakistan Peoples Party standing up to the military dictatorship of General Zia-ul-Haq. The tremendous outpouring of people from every cross section of Pakistani society—urban and rural, poor masses and middle-class businesspeople, academics, civil society leaders, students—was seen in Pakistan and all over the world as an affirmation of the forces of democracy against the brutal dictatorship that had terrorized my nation for almost a decade. Twenty-one years later, I knew, flying from Dubai to Karachi, that the size of my welcome home in 2007 would be compared to that of 1986. I knew that those around General Pervez Musharraf were chomping at the bit to proclaim that a tepid response to my return would be a legitimization of their authoritarian rule. I did not know what to expect when I stepped off that plane on October 18.

What I encountered on that day exceeded my wildest dreams and expectations. Even in an authoritarian state, literally millions of people had traveled from far and wide to greet me, and to greet the return of democracy. It was truly breathtaking, and it was covered live on Pakistani television and in real time by the BBC and CNN all over the globe. The whole world was watching, and what they saw was the rebirth of a people.

The enormity of the response did not really hit me until we left the airport on the road to Karachi. We were already far behind schedule because of the unexpectedly huge crowds at the airport itself, and the sun was setting as we crossed the Star Gate leading from the airport. As I looked down the Shar-e-Faisal highway, I saw huge numbers of people on both sides of the road, packed ten and twelve deep in a line stretching as far as the eye could see. From the beginning of the caravan, a marvelous group of brave unarmed young men in white T-shirts, volunteers who came to be known as *"Jaan Nisaar Benazir"* ("those willing to give their lives for Benazir"), surrounded my truck and held hands, making a human shield to protect me with their bodies. The security of the caravan had been a great concern of ours and a major priority for my husband as we planned my return to Pakistan. Asif had ordered bulletproof vehicles imported into the port of Karachi, but the government had denied our request. Asif had ordered jamming equipment to protect our cars and trucks from roadside improvised explosive devices (IEDs), but we were told the government would not allow us to import it. Instead we were promised that the government would provide this service itself. Having learned from the experiences in Iraq and in parts of Pakistan, we knew that these lethal devices were most often triggered by a signal from a cell phone, thus ensuring that perpetrators of terror could survive an attack by exploding their devices from a distance and then melding into the crowd.

What my husband did manage to construct to protect my safety was a raised armor-plated flatbed truck where I would be four meters off the ground and thus could be seen by the crowds even from a distance. Around the perimeter of the top of the truck where I would be standing was a four-foot-high, impenetrable bulletproof acrylic ledge that was said to be able to withstand a direct hit from even the most lethal sniper rifles. The interior of the truck was also insulated in a way that was meant to ensure that it could survive even a direct bomb attack.

The mood of the caravan was joyous. It was truly a caravan of democracy, a way for an astounding three million Pakistani citizens—including huge numbers of women and children—to come out and express their support for the PPP. Their presence was also a symbol of their support for the democratic process and their vocal opposition to the forces of dictatorship. Music pulsated from boom boxes, blasting the traditional anthems of thirty years of Peoples Party campaigns interspersed with the latest Pakistani rock music. Supporters danced around the vehicles, throwing rose petals and cheering my return and the return of democracy. People were hanging on from the trees and from telephone and electricity poles, attempting to catch a glimpse of me and the other PPP leaders who stood on the flatbed truck. It was a remarkable feeling for me after so many years abroad, years of dreaming of Pakistan, of our people, of our towns and villages, of our food and music, of the smell of basmati rice wafting from outdoor kiosks, of the sheer joy of the sound of people free and happy. I could not really believe that I was home at last and that the reception was so large and enthusiastic. The message to the world about the democratic spirit of the people of Pakistan could not have been clearer.

I kept looking around me in amazement, remembering other rallies and other campaigns. I also remembered past tragedies as well as past triumphs. As I gazed out from my truck, I saw the vibrant black, red, and green flags of the Pakistan Peoples Party everywhere, a sea of party colors. I also saw thousands upon thousands of pictures, but they weren't pictures of just me; there were huge portraits of my father, Prime Minister Zulfikar Ali Bhutto, on my left and right, in front of me and in back of me. I had an overwhelming sense that he was with me on that truck as we slowly rolled through these millions of supporters. And I also knew that the same elements of Pakistani society that had colluded to destroy my father and end democracy in Pakistan in 1977 were now arrayed against me for the same purpose exactly thirty years later. Indeed, many of the same people who had collaborated with an earlier military junta in the judicial murder of my father were now entrenched in power in the Musharraf regime and the intelligence apparatus. There could have been no more dramatic statement to me than General Musharraf's recent appointment as attorney general: Malik Qayyum, of the son of the man who had sent my father to the gallows. It was not a subtle message.

We had, of course, been discouraged from returning. Musharraf had told me in our private meetings and conversations that I should come back only after the elections. And when it was clear that I would not postpone my return, he sent messages to my staff that I should have no public demonstration or rally and I should fly directly by helicopter from the airport to Bilawal House, inside Karachi. He said that he was concerned about my security and my safety, but his supporters did very little to provide the protection we needed: jammers that worked, streetlights that worked, roads that had been cleared of empty cars that could carry improvised explosive devices (although these were not what we were expecting). This was protection to which I was entitled as a former prime minister. The general, it seemed, was quite keen to choke off any possibility that the country and the world would see the level and enthusiasm of my support. What he of course knew is that his King's Party, his private PML (Q), couldn't gather a hundred people voluntarily at a rally even at lunchtime.

I had become aware, through messages sent by General Musharraf, that suicide squads might be sent from the North-West Frontier Province and Federally Administered Tribal Areas (FATA) to try to assassinate me immediately on my return. I had actually received from a sympathetic Muslim foreign government the names and cell phone numbers of designated assassins. General Musharraf's regime knew of the specific threats against me, including the names and numbers of those who planned to kill me, and the names of others— including those in his own inner circle and in his party—who we believed were conspiring. Despite our request, we received no reports on what actions were taken before my arrival on the material provided as a follow-up to these warnings. Even as we landed, the general's people were calling to stop me from returning, stop me from giving a speech at the tomb of the Quaid-e-Azam ("The Great Leader," Mohammad Ali Jinnah), to cancel my cavalcade from the airport to the mausoleum. But I knew that those who believed in democracy and my leadership were awaiting me in the streets of Karachi and that they had come from all over the country, spending their own money and taking time off from work, to show their support for my party and for our cause of freedom and human dignity. After what they were doing for me, I thought it wrong to slip away behind their backs and skulk secretly home. I hadn't come this far in my life to abandon my people at the moment when they waited for me. And I couldn't break my word to them that I would return on October 18, 2007, return to our common hearth, home, and soil.

As the sky darkened and we progressed almost by inches through the growing masses, I began to notice a curious and disturbing phenomenon. Strangely, as we approached street corners, streetlights began to dim and then go off. After seeing that this was not an isolated incident but a clear pattern, I asked my staff to contact as many electronic and print news outlets as they could locate and inform them that streetlights were seemingly deliberately being turned off as we approached, creating a very dangerous situation. I hoped that if the situation were reported, the police authorities would take some action to keep the lights on and ensure our safety. One of my party's senators went to the utility department to press for switching the lights on. Later a supporter would tell me that a call had come to have the "lights switched off to stop the TVs giving her so much publicity." The PPP's communications secretary was able to contact at least five television outlets, and they broadcast the story almost continuously for hours. Despite our appeals, the lights were never turned back on. It was ominous. I said to a friend who was next to me, "Have you noticed the streetlights? Each one we approach goes off, so the road is in darkness and the guards can't see anything. Someone is doing this. We've had information they might try a shooting."

There was another very troubling warning that something was amiss. The jamming equipment that was supposed to be blocking cell phone signals (which could be used to detonate IEDs or suicide bombs, or even remote controlled toy planes filled with explosives) for 200 meters around my truck did not seem to be working. My husband, who had been watching the live coverage on the BBC in Dubai, called one of his friends who was with me on the truck and was helping to provide security. Asif was agitated because he was observing on television that people on my truck were talking on cell phones. He knew that if the jammers were actually working, there would be no cell phone reception anywhere near my vehicle. He demanded to know why the jammers were not working. My security advisor tried to reach Tariq Aziz, General Musharraf's National Security Council advisor, but he could not get through. Asif was concerned not only about bombs but about snipers, and he begged me to get behind the bulletproof glass and not to expose myself directly to the crowd. I said no, that I must be in front and greet my people.

I must confess I felt safe in the enormous sea of love and support that surrounded me. Moreover, I thought that the assassins would try to hit the bulletproof glass, where they expected me to be, rather than the front of the truck, where I was standing. We had been expecting attacks by snipers and by suicide bombers. We had not been expecting a car bomb. Our security was waving handheld lights, attempting to spot a suicide bomber, who would be wearing a heavy jacket or a shawl.

As midnight approached, and knowing that we were still many miles and many hours from the mausoleum, I went down into the interior of the truck with a former ambassador to the United States, Abida Hussain. Abida had been a prominent supporter of an opposition party but had recently joined the PPP with great enthusiasm. My feet had swollen up after standing in one place for ten hours, and my sandals were hurting after so much time on my feet.

I unstrapped and loosened them. A little while later my political secretary, Naheed Khan, and I went over the speech that I would be delivering later at the tomb. I considered this speech to be one of the most important of my life. This was the speech—both in substance and in symbol—that would show the world that the people of Pakistan wanted, even demanded, a transition to democracy as soon as possible, and rejected the politics of dictatorship.

As my caravan approached the neighborhood of Karsaz, the crowds had grown even larger than ever before and all around me expressed amazement. We were stunned by the overwhelming show of support, realizing that this was truly history in the making and people would forever cherish this night in the history of Pakistan. I read over the speech to Naheed, wanting to make sure that every word was right, every nuance strong and definitive. We had just finished the speech. I was saying that perhaps we should mention my petitioning the Supreme Court to allow political parties in the tribal areas to organize as part of our plan to counter extremists politically. As I said the word "extremist," a terrible explosion rocked the truck. First the sound, then the light, then the glass smashing, then the deadly silence followed by horrible screams. I knew it was a bomb. My first thought was "Oh, God, no."

When the first explosion went off exactly parallel to where I was, I physically shook with the truck, as did the others on top of the truck and inside the truck. A piercing pain tore inside my ear from the force of the blast. The heroic PPP anthems that had been pulsating from the truck suddenly stopped. An eerie silence descended that fit this moment of disbelief. Then the second explosion—much louder, larger, and more damaging—went off. Almost simultaneously with the two blasts, something hit the truck, which rolled from side to side. This was an armored truck, but it rolled back and forth. Later I saw that there were two dents clearly visible on the left side of the truck, where I was.

A European reporter and others would later tell me that sniper fire had occurred as well. Certainly the bulletproof glass where I was to have been standing had cracked where it had been pierced by either a bullet or shrapnel. The Leader of the Opposition Sindh Assembly and others would tell me of the huge orange burst, probably from a flame-thrower, that they saw go up in the forty-five-second interval between the blasts. Fire shot up around the truck. Blood and burning flesh and body parts seemed to be everywhere. Our wonderful boys in their white T-shirts holding hands and forming a human shield around my truck to protect me were the first to be mowed down. They, who only minutes before had been full of life, dancing, smiling, passing food and drinks up to the top of the truck, were now dead and dismembered. It was a massacre. It was the worst sight I had ever seen, and I'm sure the worst sight that I will ever see as long as I live.

Bodies of the dead and injured lay in the silence of the street, crumpled, the streets stained with the blood of innocents. I do not know whether the lights went on or came from the burning police-mobile and car bomb, but later on a DVD I would hear the injured say in a faint voice as life ebbed from them, *"Jeay Bhutto"*—"Long live Bhutto." Three people who were on my truck died in the blasts. The Deputy Leader of the Opposition of the Sindh Assembly received pellet and ball-bearing wounds. Others were drenched in blood and gore. I believe that the only reason we survived this assassination attempt was the dauntless courage of those young men making a human shield around the truck, thus keeping the bomb, bombers, devices, and grenades away from the base of the truck. They stopped the bombers from getting any closer to my truck and in doing so gave their lives to the cause of Pakistani democracy. Of the 179 people who died as a result of these attacks, more than 50 were these brave young men who had so much to live for. They gave their lives, leaving behind shattered loved ones.

I did not want to leave the truck guarded by the PPP security for the streets. But after eight minutes we evacuated. There was fear that the burning policemobile would trigger a fire in the truck's fuel tank. General (Ret.) Ahsan Saleem Hayat, the security chief of the procession, sent me the car he was traveling in. He was wounded, too. I was whisked away to Bilawal House, going through backstreets to evade snipers waiting for us to leave. Bilawal House is the name of my home in Karachi, which I had not seen for more than eight years. As we drove toward Bilawal House in a jeep that security boys were again clinging to, providing a human shield around me, we knew we were unarmed. We wondered whether assassins might have a backup plan to kill us, knowing we had to reach Bilawal House. This is where in 1993 Ramzi Yousef had tried to plant a bomb that he could detonate as I left home. Finally, we decided that the fortified house was the place where my safety could best be ensured.

I entered the house that my husband had built for us after our marriage, which was named in honor of our eldest child, our son, Bilawal. Going up the stairs, I saw the pictures of my three children peering back at me, and I realized the absolute terror they must be experiencing, not knowing if I were dead or alive. I had been traumatized by my father's arrest, imprisonment, and murder, and I know that such mental scars are permanent. I would have done anything to spare my children the same pain that I had undergone—and still feel—at my father's death. But this was one thing I couldn't do; I couldn't retreat from the party and the platform that I had given so much of my life to. The enormous price paid by my father, brothers, supporters, and all those who had been killed, imprisoned, and tortured, all the sacrifices made, had been for the people of Pakistan. Without realizing it, I was in shock. I saw the reality, and yet I did not see it. I thought instead of all that had to be done for the dead and for the wounded, and to find out what had happened to those on the truck with me. Despite my pleas not to expose themselves to harm on account of me, the entire leadership of the party had insisted on standing there with me shoulder to shoulder, never shirking the dangers. I spoke to my husband and assured him that I was not injured. I could not speak to my children to assure them that I was all right. Thankfully they had gone to bed and had not seen the blast on television. My daughter told me later she went to bed happy thinking of the warm reception I was getting, only to wake to a text message from a friend: "Oh, my God, I am so worried. Is your Mother all right?" With her heart pounding, she ran to the room of her father, who gathered her in his arms, reassuring her, "Your mother is fine."

It was 6:00 A.M. before I went to sleep for a few fitful hours, only to wake to a bitter reality. As the plotters had probably intended, the story of the massacre had replaced the story of the three million who had greeted me on my return. Despite the fact that the hospitals of Karachi were overflowing with eyewitnesses to the assassination attempt, the police were apparently not conducting an investigation. There were no forensic teams collecting evidence at the bombing site, and with every passing minute, potentially critical evidence was disappearing from the scene. Instead of the site being cordoned off to protect evidence, it was scrubbed clean within hours and the evidence was destroyed. No one from the police or the government was collecting testimony from the victims of the attack. A cover-up seemed to be under way from the very first moments of the attack. The provincial government announced that it had been a suicide attack.

Clearly this was meant to appear to be an Al Qaeda–style suicide attack, more Muslim-against-Muslim violence linked to the so-called struggle between theology and democracy. But in Pakistan things are almost never as they seem. There are always circles within circles, rarely straight lines. This was meant to look like the work of Al Qaeda and the Taliban, and I do not doubt that they were involved. But the sophistication of the plan—the multiple explosions, the flamethrower, the gas in the air, the dents in the truck, and the unreported fact that snipers had fired high-powered long-distance bullets—suggested a larger conspiracy. Elements from within the Pakistani intelligence service had actually created the Taliban in the 1980s, and certain elements sympathized with Al Qaeda ideologically and theologically. Some had recruited for or worked with it. I had identified those I suspected in my letter to the general before my return.

At first the government put someone in charge of the investigation who had actually been involved in the torture and near death of my husband, Asif, in 1999. Adding insult to the gravest injury, the president's ruling PML clique, playing its usual blame-the-victim game, publicly claimed that the attack had been staged by the PPP to generate sympathy.

The interior minister, who had crossed over from my party and joined Musharraf in 2002, and who thus had a vested interest in the political status quo, assumed overall responsibility for the investigation. I publicly demanded that the FBI's and Scotland Yard's forensic teams—universally considered the best in the world—be brought in to assist in the investigation. Minister Aftab Sherpao immediately refused, claiming it was a violation of Pakistani sovereignty. Of course, there were several precedents for seeking outside technical help on investigations. Pakistan had asked the FBI for assistance in the investigation of the bombing of the Egyptian Embassy in Islamabad. It had also asked for FBI forensic assistance in the investigation of the death of then–Army Chief of Staff Asif Nawaz in 1993. I asked international detectives for assistance in the investigation of the death of my brother Murtaza in 1996. And both the FBI and Scotland Yard were brought in to do forensic analysis on the plane on which General Zia-ul-Haq and Ambassador Arnie Raphel had died in 1988. There was one precedent after another, spanning three different Pakistani administrations, which would have allowed for forensic assistance in the investigation of the mass murders of the early hours of October 19. Yet the Musharraf military regime refused. If there were nothing to hide, I reflected, why the insecurity and refusal to facilitate the investigation?

To this day, I have not been asked to give testimony on the events of that night.

The terrorist assassination attempt against me on October 19, 2007, underscores the issues troubling me about internal strife within the Islamic fold and the intersection of Islam and democracy. The Muslim-on-Muslim massacre that took place in Karachi in October is consistent with the Muslim-on-Muslim fratricidal sectarian violence that is raging in Baghdad and other parts of Iraq in the early twenty-first century. The Talibanization of the Federally Administered Tribal Areas of Pakistan and the growth of extremism within the North-West Frontier Province of my country highlight my central concern and the reason for writing this book. The potential exists for the radicalization of Muslims around the world in a political environment of dictatorship and authoritarianism. If extremism and militancy thrive under dictatorship and cannot be contained by a one-man show relying on military might, the democratic world would have a strong rationale—if not for moral reasons then at least for reasons of self-interest—in helping to sustain democratic governance in the nations of the Muslim world.

It is with this in mind that I will recall the record and history of the West in promoting or discouraging the growth of democratic institutions through three centuries in predominantly Muslim nations. I will argue that the fundamentals of democratic governance are part of the Islamic value system and debunk the myth that Islam and democracy are mutually exclusive. I know from my own experience that democracy is an integral part of Islam. The core of my being as a Muslim rejects those using Islam to justify acts of terror to pervert, manipulate, and exploit religion for their own political agenda. Their actions are not only antithetical to Islam but specifically prohibited by it.

The central message I would like to convey in this book is of the two critical tensions that must be reconciled to prevent the clash of civilizations that some believe looms before us. There is an internal tension within Muslim society, too. The failure to resolve that tension peacefully and rationally threatens to degenerate into a collision course of values spilling into a clash between Islam and the West. It is finding a solution to this internal debate within Islam—about democracy, about human rights, about the role of women in society, about respect for other religions and cultures, about technology and modernity—that will shape future relations between Islam and the West. But both clashes can be solved. What is required is accommodation and reconciliation.

To that end, I make a small and humble contribution in writing this book to share a modern Muslim woman's view.

2

The Battle Within Islam:
Democracy Versus Dictatorship,
Moderation Versus Extremism

When Al Qaeda hijacked airplanes to attack the United States on September 11, 2001, it tried to hijack the message of my religion—the religion of Islam—as well. In doing so it ignited the great battle (which some have called a global war) of the new millennium. The murder of almost three thousand innocent people in the name of jihad is not only antithetical to the values of the civilized world but contradictory to the precepts of Islam. The terrorists exploited images of savagery and brutality for political advantage, just as demagogues before them manipulated Islam for political gain.

Adopting the philosophy of the dialectic—that to effect change, things must first dramatically deteriorate—Al Qaeda desperately tried to provoke the notorious clash of civilizations that had been prophesized years before. In doing so it twisted the values of a great and noble religion and potentially set the hopes and dreams of a better life for Muslims back a generation. The damage was not limited to New York, Washington, and Pennsylvania. Muslims, and the Muslim world, became their victims, too.

It is the tradition of Islam that has allowed me to battle for political and human rights, and this same tradition strengthens me today. Islam denounces inequality as the greatest form of injustice. It enjoins its followers to combat oppression and tyranny. It enshrines piety as the sole criterion for judging humankind. It shuns race, color, and gender as the basis of distinctions within society.

Islam is committed not only to tolerance and equality but to the principles of democracy. The Quran says that Islamic society is contingent on "mutual advice through mutual discussions on an equal footing." Islam condones neither cruelty nor dictatorship. Beating, torturing, and humiliating women are inconsistent with its principles. Denying education to girls violates the very first word of the Holy Book: "Read." According to our religion, those who commit cruel acts are condemned to destruction.

Islam is not the caricature that is often portrayed in Western media. Rather, it is an open, pluralistic, and tolerant religion—a positive force in the lives of more than one billion people across this planet, including millions in the growing Islamic populations of Europe and the United States. It is a religion built upon the democratic principles of consultation (*shura*); building consensus (*ijma*); finally leading to independent judgment (*ijtihad*). These are also the elements and processes of democratic institutions and democratic governance.

During the darkness of the Middle Ages in northern Europe, when barbaric hordes raped and pillaged at will, Islam was building the great libraries and universities of the world, developing the arts, sciences, and humanities. When women were viewed as inferior members of the human family and treated as property belonging to men all over the globe, the Prophet Mohammad accepted women as equal partners in society, in business, and even in war. Islam codified the rights of women. The Quran elevates the status of women to that of men. It guarantees women civil, economic, and political rights.

Later to become the Prophet's wife, Bibi Khadijah, the rich and successful businesswoman, hired the Prophet Mohammad when she heard of his reputation for honesty and his upstanding nature. When the Prophet Mohammad received his first revelation from God and had doubts and fears, it was Bibi Khadijah who believed that his experience truly was divine; she comforted and encouraged him. She became the world's first Muslim. The world's first person to embrace Islam was a woman!

Throughout the Holy Quran, there is example after example of respect for women as leaders and acknowledgment of women as equals. Again, the first word of the Holy Book is "Read." It does not say, "Men Read"; it says, "Read." It is a command to all believers, not just to men. For in the religion of Islam in which I was brought up, there is only equality.

Those who would pervert Islam by committing crimes against the innocent have continued their war on civilization since the attacks on America. They have attacked the innocent in Madrid, in London, in Riyadh, and in Bali. And the extremists and fanatics struck against my people and me on October 19, 2007, in Karachi, staining the streets of my hometown with the blood of 179 martyrs to democracy, most of whom were supporters of the Pakistan Peoples Party and party security guards who lost their lives protecting me.

The perpetrators of these crimes against humanity are those the Quran describes as "going astray from the right path." There are those who claim to speak for Islam who denigrate democracy and human rights, arguing that these values are Western values and thus inconsistent with Islam. These are the same people who would deny basic education to girls, blatantly discriminate against women and minorities, ridicule other cultures and religions, rant against science and technology, and endorse brutal totalitarianism to enforce their medieval views. These people have no more legitimate relationship to Islam than the people who bomb women's health centers in America have to Christianity or the madmen who massacre innocent Arab children at the tomb of Abraham in Palestine have to Judaism. Throughout history, the greatest crimes against humanity have been those carried out in the name of God, fanaticizing religious values to justify unspeakable acts against civilization.

The battle for the hearts and soul of Islam today is taking place between moderates and fanatics, between democrats and dictators, between those who live in the past and those who adapt to the present and plan for a better future. In the resolution of this conflict may in fact lie the direction of international peace in the twenty-first century. For if the fanatics and extremists prevail—if those who attacked America, Spain, Britain, Indonesia, Afghanistan, and Pakistan and then attacked my supporters and me in Karachi on October 19, 2007, succeed—then a great *fitna* (disorder through schism or division) could sweep the world. Here lies their ultimate goal: chaos.

The case—the very strong case—for a pluralistic and modern Islamic society is made directly in Islamic scriptures' references to violence, terrorism, intercultural relations, interreligious relations, the place of women in society, and science and technology. Despite the protestations and assertions of some, and despite skepticism outside our own community, the vast majority of the billion Muslims in the world embrace a peaceful, tolerant, open, rational, and loving religion that codifies democratic values. It is a religion that sanctifies the traditions of the past while embracing the hope for progress in the future. This is the interpretation of Islam that my father, Zulfikar Ali Bhutto, and my mother, Nusrat Bhutto, embraced and practiced and taught my brothers, Mir Murtaza and Shah Nawaz, my sister, Sanam, and me. This is the true Islam, in contrast to the perversion that has been espoused by extremists and militants and the caricature that is too often accepted in the West. The greatest and purest source is the words of the Prophet himself. And when the Prophet speaks of "Allah," he is speaking of God, the same monolithic God of Judaism, Christianity, and Islam. "God" is a translation of the Arabic word "Allah," not just the God of Islam but rather the God of monotheism, the God of all who believe in Him and believe that He is the Creator of the Universe, of this world and the hereafter.

·

I believe there is great confusion around the world about whether violence is a central precept of Islam because of a basic misunderstanding of the meaning of the term "jihad." Because terrorists call their murderous acts jihad, much of the world has actually come to believe that terrorism is part of an ordained, holy war of Islam against the rest of humanity. This perception must be dispelled immediately.

Many people around the world think that the word "jihad" means only military war, but this is not the case. As a child I was taught that jihad means struggle. Asma Afsaruddin, a well-regarded scholar of Islam, explains the correct meaning well: "The simplistic translation of *jihad* into English as 'holy war,' as is common in some scholarly and nonscholarly discourses, constitutes a severe misrepresentation and misunderstanding of its Quranic usage." Jihad instead is the struggle to follow the right path, the "basic endeavor of enjoining what is right and forbidding what is wrong."

The importance of jihad is rooted in the Quran's command to struggle (the literal meaning of the word "jihad") in the path of God and in the example of the Prophet Mohammad and his early companions. In fact, some in the Muslim world believe holy war against the West is warranted in a new global conflict.

Clearly there are some Muslims today who believe that the conditions of their world require a jihad. They look around them and see a world dominated by corrupt authoritarian governments and a wealthy elite, a minority concerned solely with its own economic prosperity rather than national development. They see a world awash in Western-dominated culture and values in dress, music, television, and movies. "Western governments are perceived as propping up oppressive regimes and exploiting the [Islamic world's] human and natural resources, robbing Muslims of their culture and their options to be governed according to their own choice and to live in a more just society."

A small, violent minority of Muslims associated with the defensive Afghan jihad of the 1980s against the Soviet occupation of Afghanistan believe they defeated one superpower and can defeat another. They plan to mobilize an offensive "holy" army to fight the West in either Afghanistan or parts of Pakistan using terrorist attacks against Muslim and non-Muslim civilians, which will somehow liberate

Muslims everywhere from the yoke of decadence and Western domination. A discussion of jihad is critical to the world and critical to this book. If jihad is indeed about offensive holy war against other religions and Muslim sects, then surely Muslims will have trouble living in a democratic world, let alone forming their own functioning, pluralistic democracies.

If this is the case, then my thesis—that democracy and Islam are not only compatible but mutually sustaining—will fail. Therefore, it is important to show the true meaning of jihad: as an internal and external struggle to follow the right and just path. Jihad involving armed conflict must be constrained by the standards of just war, just as Christianity sets similar standards. I will substantiate, with theological backing, the idea that terrorism cannot be supported by reference to the Holy Book.

In the history of Islam, there are two different constructs of the term "jihad." First there is the internal jihad, a jihad within oneself to be a better person, to resist the temptations of the soul. This is a struggle centered on eradicating character flaws such as narcissism, greed, and wickedness. This is the greater jihad. The second form of jihad is personal conduct at a time of war or conflict. The Prophet is said to have remarked when he came home from a battle, "We return from the lesser jihad to the greater jihad." This shows the importance of the constant internal struggle that we all face within ourselves. It is nonviolent struggle that makes us become better people. The greater, internal jihad is seen as more important than the lesser, external jihad.

The concept of the lesser jihad is mentioned many times throughout the Quran, and the Quran gives it multiple meanings. A closer look at the Quran will show that the concept evolved according to the context in which the suras (chapters of the Quran) were revealed. When the Prophet began revealing God's word in Mecca, it was a city of violence, of tribal warfare and personal vendetta. It was a city based exclusively on tribal allegiance. The primary religion in Mecca was paganism. Thus, when the Prophet began preaching the Quran, he brought forth new ideas and a universal system of laws that surpassed the traditional tribal law, which he rejected. His new community of Muslims became a persecuted minority that could not assert itself, that was oppressed and attacked. The quranic revelations received at this time reflect the period in which they were revealed. Thus, the definition of jihad that emerged is almost exclusively defensive and nonviolent. As is written in the Quran: "And the recompense of evil is punishment like it, but whoever forgives and amends, he shall have his reward from Allah; surely He does not love the unjust. And whoever defends himself after his being oppressed, these it is against whom there is no way (to blame)." The Holy Book continues: "And whoever is patient and forgiving, these most surely are actions due to courage." Jihad seems to be limited to defensive fighting only, and even in defense, the above verse seems to preach a nonviolent solution over a violent solution when possible.

In A.D. 622 the Prophet Mohammad and his community of Muslims left the persecution of Mecca and established the world's first Islamic political system in Medina. From this new position they fought three wars with the Meccans and many other skirmishes. The revelations of the Quran that were given to Mohammad during this time delineate the criteria for a permissible, or "just," war. This first verse reflects the more violent period in which it was received: "Permission (to fight) is given to those upon whom war is made because they are oppressed, and most surely Allah is well able to assist them." Clearly, war is justified when it is a defensive war. Additionally, this next verse is explicitly against aggressive warfare: "And fight in the way of Allah with those who fight with you, and do not exceed the limits, surely Allah does not love those who exceed the limits."

Some of the verses dealing with war and violence during this time period seem to be more lenient as to when violence would be permitted and condoned. They are referred to as the "sword verses":

So when the sacred months have passed away, then slay the idolaters wherever you find them, and take them captives and besiege them and

*lie in wait for them in every ambush, then if they repent and keep up
prayer and pay the poor rate, leave their way free to them; surely Allah
is Forgiving, Merciful.*

At first glance this verse seems to advocate violence against unbelievers. But its context is a specific battle in Medina occurring at the
time of its revelation, a battle against idol worshippers, not people of
the Book, not believers in monotheism. It commands that violence
cease if the offenders repent. The second sword verse is as follows:

*Fight those who do not believe in Allah, nor in the latter day, nor do
they prohibit what Allah and His Messenger have prohibited, nor follow the religion of truth, out of those who have been given the Book,
until they pay the tax in acknowledgment of superiority and they are in
a state of subjection.*

Although this verse may appear superficially problematical, a close
reading shows that it does not advocate violence against people of
the Book, only those who reject God and his teachings outright. (Let
us always be sensitive to the fact that the word "Allah" is simply the
Arabic translation of the English word "God" or the Hebrew word
"Jehovah.") And as a later verse (2:193) shows, the offenders should be
fought only until they cease hostilities toward Muslims, implying
that those not initiating hostilities cannot be targeted. And last, if an
enemy requests peace, it must be given: "And if they incline to peace,
then incline to it and trust in Allah; surely He is the Hearing, the
Knowing."

A contemporary scholar, Majid M. Khadduri, gives a good explanation of *bellum justum* ("just war") in Islamic tradition. Jihad as just
war is defensive by nature. However, as in the Christian and Roman
traditions, there are certain other justifications that can be used for
going to war. According to Khadduri, changes were made in Islamic
bellum justum theory as time went by. There was often a need by Muslim states to make peace, and not on their own terms. Therefore,

Muslim jurists began to reinterpret law and to justify the suspension of jihad. They agreed on the necessity of peace. Some said that jihad as permanent, external struggle was now obsolete and no longer compatible with Muslim interests; this did not mean the abandonment of jihad duty, just its suspension.

This shift in the conception of jihad from active to dormant also reflected the end of the territorial expansion of Islam and the revival of intellectual and philosophical Islam (ca. A.D. ninth century). For thinkers like Ibn Khaldun the change in the character of the state from a warlike one to a peaceful one meant movement toward a civilized state. As Khadduri explains, "In both Islam and ancient Rome, not only was war to be *justum* but also to be *pium* [holy], that is, in accordance with the sanction of religion and the implied commands of gods."

Yet jihad is not one of the Five Pillars of Islam (except in Khariji theory), reflecting the theological level of importance attached to just war. War, as in many religious traditions, can be justified in Islam when necessary but is not seen as a continuous religious duty, as some uninformed Western sources would have one think. Islam was able to regulate war in a region that experienced constant warfare. War existed in pre-Islamic Arabia, but, as Khadduri notes, "Islam outlawed all forms of war except the *jihad,* that is, the war in *Allah's* path."

Additionally, external jihad as an individual duty is also regulated. Khadduri argues that if one part of the community participates in justified external jihad when necessary, the rest of the community is absolved of its responsibility to participate in the external jihad. This of course shows the Islamic basis for external jihad as an instrument of the state. Regulated armies seem to be preferred to irregular elements of society participating in jihad in an unorganized (and usually ineffective) manner:

> Jihad *is a collective obligation of the whole Muslim community* (fardl-kifaya). . . . *If the duty is fulfilled by a part of the community it ceases to be obligatory on the others; the whole community, however, falls into*

error if the duty is not performed at all. Imposition of jihad *on the community rather than on the individual is very significant and involved at least two important implications. In the first place, it meant that the duty need not necessarily be fulfilled by all the believers. . . . In the second place, the imposition of the obligation on the community rather than on the individual made possible the employment of* jihad *as a community and, consequently, a state instrument.*

Certainly, the Quran and the hadith argue for dying for a just cause. Two hadith examples are illustrative. Muslim ibn al-Hajjaj (d. 875 C.E.) and Ibn Maja (d. 887) gave reports that claim that God forgives martyrs for all sin but debt. Abd al-Razzaq al-San'ani (d. 826) quotes the Prophet: "When one of you stands within the battle ranks, then that is better than the worship of a man for sixty years." These verses do support God's forgiveness for those who die in just causes. However, later jurists and extremists who allege that the Quran supports the actions of terrorists who take their life to kill innocents do not have textual support. Suicide-murder is specifically and unambiguously prohibited in the Holy Book: On that account:

For this reason did We prescribe to the children of Israel that whoever slays a soul, unless it be for manslaughter or for mischief in the land, it is as though he slew all men; and whoever keeps it alive, it is as though he kept alive all men; and certainly Our messengers came to them with clear arguments, but even after that many of them certainly act extravagantly in the land.

Thus, in the Quran, preserving life is a central moral value. The Quran once again shows God's preference for life over death in this next verse: "He who disbelieves in Allah after his having believed, not he who is compelled while his heart is at rest on account of faith, but he who opens (his) breast to disbelief—on these is the wrath of Allah, and they shall have a grievous chastisement."

The Quran holds saving one's life in such high regard that it allows one to renounce his faith if he is under duress, as long he keeps his true faith in his heart (that is, he does not actually renounce it).

These verses demonstrate the value the Quran puts on life; it does not permit suicide but demands the preservation of life: "And spend in the way of Allah and cast not yourselves to perdition with your own hands, and do good (to others); surely Allah loves the doers of good." The Holy Book goes on to give another specific prohibition of suicide (although on the group level): "O you who believe! do not devour your property among yourselves falsely, except that it be trading by your mutual consent; and do not kill your people; surely Allah is Merciful to you." The Quran is thus explicit in denying the validity of murder-suicide in its teachings.

Let us look specifically at the issue of terrorism. Muslim jurists developed a specific body of laws called *siyar* that interprets and analyzes the just causes for war. Part of the law indicates that "those who unilaterally and thus illegally declare a call to war, attack unarmed civilians and recklessly destroy property are in flagrant violation of the Islamic juristic conception of *bellum justum*. Islamic law has a name for such rogue militants, *muharibun*. A modern definition of *muharibun* would very closely parallel the contemporary meaning of 'terrorists.' The acts that these *muharibun* commit would be called *hiraba* ('terrorism'). Thus all terrorism is wrong. There is no 'good terrorism' and 'bad terrorism.' " Osama bin Laden's creed that "the terrorism we practice is of the commendable kind" is an invented rationalization for murder and mayhem. In Islam, no terrorism—the reckless slaughter of innocents—is ever justified.

Given the importance of the concept of terrorism to this book, it is essential to review how some of the most influential thinkers in the Islamic reform movement have dealt with this issue. The Islamic reform movement is the term used to describe the group of Muslims in the last two centuries that have sought to reformulate some Muslim ideologies by following a literalist, more text-based interpretation of Islam. These are the sources to which many extremists look for guidance and support. The medieval scholar Ahmad Ibn Taymiyya advocated a return to the ideals of the first Muslim community at Medina. He drew a sharp line between Muslims and nonbelievers and asserted that "Muslim citizens thus have the right, indeed duty, to revolt against them [nonbelievers], to wage *jihad.*" His dictum would be copied by many groups, such as the Egyptian Islamic Jihad (the assassins of former Egyptian president Anwar Sadat) and Osama bin Laden's Al Qaeda. They make a sharp distinction between what they see as truth and untruth, and those who do not believe in their truth are branded as unbelievers; in their eyes, a call for jihad is thus warranted against those unbelievers.

Maulana Maudoodi, the founder of the extremist group Jamaat-i-Islami (JI) in South Asia, believed that Muslim identity was threatened by the rise of nationalism in South Asia during the first half of the twentieth century. He saw nationalism as a Western ideology unilaterally imposed upon Muslims to weaken and divide the community by replacing the idea of a worldwide Muslim community with individual nationalisms based on language, ethnicity, and locality. He believed that Islam can overcome these obstacles, "and so Maudoodi sees Muslims as an international party organized to implement Islam's revolutionary program, and *jihad* as the term that denotes the utmost struggle to bring about an Islamic revolution."

One of the strongest intellectual forces behind Islamic extremism was Sayyid Qutb, a twentieth-century Muslim activist from Egypt's Muslim Brotherhood. He used the term *"jahiliyyah"* (the term used in the Quran for the pre-Islamic world, a period of ignorance) to describe the modern world. He was disgusted with the culture of the West and the dictatorial governments of the Muslim world. He saw the West as a historical enemy of Islam (as exemplified by the Crusades, colonialism, and the Cold War) and also saw the elite ruling class of Muslims as corrupt. Given the authoritarian nature of most governments in the Islamic world, Qutb believed that minor changes within the current systems in the Islamic world were insufficient. Instead he believed and proposed that offensive, violent jihad was the only way to impose his view, that of a new Islamic world community, on the world.

These three reactionaries represent a type of thinking currently popular in parts of the Islamic world. The adherents to these views want to see a return to what they claim to be the fundamentals of Islam. They see the West as corrupting Islamic countries with the collusion of Muslim elites. Using mistaken interpretations of the Quran, they believe that they can justify acts of violence against innocents, people of the Book, and even fellow Muslims in order to achieve their goals. Clearly, the Quran does not support the teachings of these reactionary clerics. They may provide an intellectual infrastructure for the terrorist movement, but it is a house of false cards and twisted logic. Fanatics will use every rationalization to justify their terror, and this has traditionally been true for religious extremists.

Let us look at the most infamous contemporary terrorist, Osama bin Laden, to prove this point. In a 1998 interview, he claimed: "We do not have to differentiate between military or civilian. As far as we are concerned, they are all targets." Bin Laden unilaterally violated the principles laid down by the Quran to serve his own narrow-minded political ends. As we saw in the quranic text, the Quran places high value on the sanctity of human life and permits the use of violence only in extreme situations, such as in the defense of one's community against invaders. Bin Laden's utter disregard for the value of human life, especially his doctrine of including innocents in the senseless carnage, is un-Islamic. Indeed, in addition to putting value on all human life, the Quran puts a special value on the lives of innocents: "So they went on until, when they met a boy, he slew him. (Musa) said: Have you slain an innocent person otherwise than for manslaughter? Certainly you have done an evil thing."

Bin Laden is not representative of Islam, or any civilization, for that matter. Of course, this problem of religious fanatics hijacking religious values to serve their own violent interests is not a problem limited to Islam. I have long compared bin Laden's attempt to exploit, manipulate, and militarize Islam to terrorist acts by other religious fanatics—whether Christian fundamentalists' attacks on women's reproductive clinics or Jewish fundamentalist attacks on Muslim holy sites in Palestine. All the children of the Book have suffered from those who would use force in God's name to achieve political objectives.

•

Some may question whether Muslims can use ancient texts for explanation and guidance in the modern world. But certainly those who are followers of any religion accept the universality of its respective doctrine. The Old Testament, the New Testament, and the Quran were not texts meant only for the times of their revelation but texts for all time, meant to guide through the ages.

But some would question whether looking at the Quran or any other Abrahamic text now (in the modern world) for ideas such as pluralism and individual autonomy is merely forcing modernist notions of human rights and other democratic ideals on a message revealed in another era. For issues not explicitly discussed in the Quran, one turns to modern religious jurists for interpretation. These scholars can evaluate the historical context at the time the Quran was received and interpolate universal principles that can be applied to contemporary issues. The message of Islam is subject to *ijtihad* and *ijma.* In every age, reason is applied to its constant principles to arrive at a consensus of interpretation for that age.

There has been a raging debate within Islam on how Islam relates to other cultures and other religions. Islam has actually embraced other cultures and religions in ways much more accepting and respectful than any of the other great monotheistic religions of our time in their early periods. Islam may now have the image of being closed and intolerant, but nothing could be further from the truth, as much as extremists would like the world to think otherwise.

Islam accepts as a fundamental principle the fact that humans were created into different societies and religions, and that they will remain different: "And if your Lord had pleased He would certainly have made people a single nation, and they shall continue to differ." And: "And if your Lord had pleased, surely all those who are in the earth would have believed, all of them; will you then force men till they become believers?" God did not will everyone on earth to be adherents of one religion or members of one culture. If He had wanted this, He would have ordained it so. This means that God created diversity and asked believers to be just and to desire justice in the world. Thus it flows that God wants tolerance of other religions and cultures, which are also created by him.

The Quran reveals that God sent 120,000 prophets. Thus, it can be argued that in a Muslim state, diverse points of view will be represented and must be protected. The Quran does not simply preach tolerance of other religions; it also acknowledges that salvation can be achieved in all monotheistic religions. Freedom of choice, especially in matters of faith, is a cornerstone of quranic values. This freedom, of course, leads to pluralism in religion, both within Islam and outside. The quranic preference for freedom of choice clearly manifests a divine desire for pluralism and religious diversity; examples of this from the Quran are clear and striking: "You shall have your religion and I shall have my religion." The Quran unambiguously desires choice in religious matters.

Quite remarkably and uniquely, the Quran acknowledges that other religions can readily lead to salvation. For example, the Holy Book says:

Surely those who believe, and those who are Jews, and the Christians, and the Sabians, whoever believes in Allah and the Last Day and does good, they shall have their reward from their Lord, and there is no fear for them, nor shall they grieve.

Islam embraces all humanity under one God, discrediting all other exclusive religious claims to salvation. I don't believe there is anything quite like this in any religion on earth.

The Quran promotes religious pluralism. It does not seek to cancel out or supersede previous revelations in the form of religions such as Judaism and Christianity. Instead, the multitude of monotheistic religions is seen by God as serving a purpose: the establishment of morally upright and ethical people. God created man with intrinsic values of justice and equality. This global community that God created is commanded to "strive with one another to hasten to virtuous deeds" or, in other translations, "compete with one another in good works."

The Quran specifically sanctifies those who believe in the one true God and live a good and virtuous life. It does not say that only Islam is the route to salvation. No human being can limit divine mercy in any way. Thus we on earth cannot differentiate between religions. Examples from the Quran:

> *Whatever Allah grants to men of (His) mercy, there is none to withhold it, and what He withholds there is none to send it forth after that, and He is the Mighty, the Wise. . . .*
>
> *Those who bear the power and those around Him celebrate the praise of their Lord and believe in Him and ask protection for those who believe: Our Lord! Thou embracest all things in mercy and knowledge, therefore grant protection to those who turn (to Thee) and follow Thy way, and save them from the punishment of Hell. . . .*
>
> *Will they distribute the mercy of your Lord? We distribute among them their livelihood in the life of this world, and We have exalted some of them above others in degrees, that some of them may take others in subjection; and the mercy of your Lord is better than what they amass.*

Islam believes that people should be allowed choice in religion and that no religion should be forced upon people. Religious freedom allows people to freely accept Islam—or any religion—as their religion of choice. Practicing Islam out of force rather than choice is un-Islamic. The Quran states: "There is no compulsion in religion." Indeed, the practice of religion while under compulsion or forced conversion is against the teachings of the Quran. (Thus, the forced conversion of Muslims perpetrated by Christian slave masters would not have been permitted in Islam, nor would have the forced conversion of Sephardic Jews by Christians during the Spanish Inquisition.)

Islamic teachings, through the Holy Quran, encourage the principle of pluralism. The Quran emphasizes coexistence with others. It is God who has created the universe into many tribes and nations. All are God's creatures, and according to the Quran, each man and woman created is equal before the eyes of God. It is this equality, irrespective of race or religion, that underlies the pluralism and tolerance at the heart of Islam: "If Allah had pleased He would have made you (all) a single people, but that He might try you in what He gave you, therefore strive with one another to hasten to virtuous deeds."

Human beings have a common ethical responsibility toward one another, having been created from one soul (Adam and Eve were created from one soul, and we are all their descendants): "O people! be careful of (your duty to) your Lord, Who created you from a single being and created its mate of the same (kind) and spread from these two, many men and women."

One of the most important messages of the Quran is that there is no compulsion in the way we live our lives. This lays the basis of tolerance in human conduct. All the great Muslim leaders respected non-Muslims and did not forcibly convert or kill people on the basis of their religious belief. The principle "You shall have your religion and I shall have mine" has been incorporated into what we now call "the right to religious freedom." This teaching prohibiting compulsion gives substance to Islam's belief in freedom of expression and lays the basis for debate and discussion, the essential traits of a democratic society.

There is a historical precedent for different religious communities living peacefully together under Muslim rule, even in the very first Muslim community under the leadership of the Prophet Mohammad. In A.D. 622 the Prophet migrated from Mecca to Medina, where he established the first Islamic state. He based this state on a compact among three communities: Mahajirun, Ansar, and Yahud. This state was made up of different religions, yet this early contract established a simple federal state, which gave all three communities equal rights.

Islam is actually quite extraordinary, among the three great monotheistic religions, in accepting earlier religions' messages sent through Moses and Christ as its own. In fact, *"isla"* means "submission before the one and only God." *"Muslim"* means "one who submits before the one and only God." Christians and Jews also believe in one God. Therefore, when the British Empire was at its zenith and the sun never set on it, the British decided to call Muslims "Mohammedans," followers of the Prophet Mohammad. This was done to differentiate "Mohammedans" from the followers of Christ and Moses, who also by virtue of worshipping one god could have been described as "Muslim."

In reading the Quran, Christians and Jews will see echoes of their own religious teachings in certain passages. The extraordinary commonalities of the world's three great monotheistic religions should promote tolerance among them.

And the same did Ibrahim {Abraham} enjoin on his sons and (so did) Yaqoub {Jacob}. O my sons! surely Allah has chosen for you (this) faith, therefore die not unless you are Muslims. [Muslims meaning Believers in one God.]

Nay! were you witnesses when death visited Yaqoub {Jacob}, when he said to his sons: What will you serve after me? They said: We will serve your God and the God of your fathers, Ibrahim {Abraham} and Ismail and Ishaq {Isaac}, one God only, and to Him do we submit. . . .

Surely Allah chose Adam and Nuh {Noah} and the descendants of Ibrahim {Abraham} and the descendants of Imran above the nations. . . .

Say: O followers of the Book! come to an equitable proposition between us and you that we shall not serve any but God and (that) we shall not associate aught with Him, and (that) some of us shall not take

*others for lords besides God; but if they turn back, then say: Bear wit-
ness that we are Muslims [one who submits to one God]. . . .*

*Say: We believe in Allah and what has been revealed to us, and
what was revealed to Ibrahim {Abraham} and Ismail and Ishaq
{Isaac} and Yaqoub {Jacob} and the tribes, and what was given to
Musa {Moses} and Isa {Jesus} and to the prophets from their Lord; we
do not make any distinction between any of them, and to Him do we
submit. . . .*

*Say: Surely, (as for) me, my Lord has guided me to the right path;
(to) a most right religion, the faith of Ibrahim the upright one, and he
was not of the polytheists. . . .*

*And mention Ibrahim in the Book; surely he was a truthful man, a
prophet. . . .*

*Surely We revealed the Taurat {Torah, the Jewish Holy Text} in
which was guidance and light; with it the prophets who submitted
themselves (to God) judged (matters) for those who were Jews, and the
masters of Divine knowledge and the doctors, because they were required to guard (part) of the Book of God, and they were witnesses
thereof; therefore fear not the people and fear Me, and do not take a
small price for My communications; and whoever did not judge by what
God revealed, those are they that are the unbelievers. . . .*

*Again, We gave the Book to Musa {Moses} to complete (Our bless-
ings) on him who would do good (to others), and making plain all
things and a guidance and a mercy, so that they should believe in the
meeting of their Lord. . . .*

*And most certainly We gave Musa {Moses} the Book and We sent
messengers after him one after another; and We gave Isa {Jesus}, the son
of Marium {Mary}, clear arguments and strengthened him with the
holy spirit, What! whenever then a messenger came to you with that
which your souls did not desire, you were insolent so you called some li-
ars and some you slew. . . .*

Say: We believe in God and (in) that which had been revealed to us, and (in) that which was revealed to Ibrahim {Abraham} and Ismail and Ishaq {Isaac} and Yaqoub {Jacob} and the tribes, and (in) that which was given to Musa {Moses} and Isa {Jesus}, and (in) that which was given to the prophets from their Lord, we do not make any distinction between any of them, and to Him do we submit. . . .

He has revealed to you the Book with truth, verifying that which is before it, and He revealed the Taurat {Torah} and the Injeel {Gospel} aforetime, a guidance for the people. . . .

Surely We have revealed to you as We revealed to Nuh {Noah}, and the prophets after him, and We revealed to Ibrahim {Abraham} and Ismail and Ishaq {Isaac} and Yaqoub {Jacob} and the tribes, and Isa {Jesus} and Ayub {Job} and Yunus {Jonah} and Haroun {Aaron} and Sulaiman {Solomon} and We gave to Dawood {David}. . . .

And We sent after them in their footsteps Isa {Jesus}, son of Marium {Mary}, verifying what was before him of the Taurat {Torah} and We gave him the Injeel {Gospel} in which was guidance and light, and verifying what was before it of Taurat {Torah} and a guidance and an admonition for those who guard (against evil). . . .

And when We made a covenant with the prophets and with you, and with Nuh {Noah} and Ibrahim {Abraham} and Musa {Moses} and Isa {Jesus}, son of Marium {Mary}, and We made with them a strong covenant.

Muslims believe that the Quran is the continuation of the message sent earlier through other prophets, including the prophets of Judaism and Christianity. This is in contrast to other world religions, which "break" from past religions. For example, the New Testament says about Jews: "You belong to your father, the devil, and you want to carry out your father's desire."

Woe to you, teachers of the law and Pharisees, you hypocrites! You build tombs for the prophets and decorate the graves of the righteous. And you say, "If we had lived in the days of our forefathers, we would not have taken part with them in shedding the blood of the prophets. So you testify against yourselves that you are the descendants of those who murdered the prophets. Fill up, then, the measure of the sin of your fore-fathers! You snakes! You brood of vipers! How will you escape being condemned to hell? Therefore I am sending you prophets and wise men and teachers. Some of them you will kill and crucify; others you will flog in your synagogues and pursue from town to town." . . .

The word of the Lord spread through the whole region. But the Jews incited the God-fearing women of high standing and the leading men of the city. They stirred up persecution against Paul and Barnabas, and expelled them from their region. . . .

Men of Israel, why does this surprise you? Why do you stare at us as if by our own power or godliness we had made this man walk? The God of Abraham, Isaac and Jacob, the God of our fathers, has glori-fied his servant Jesus. You handed him over to be killed, and you dis-owned him before Pilate, though he had decided to let him go. You disowned the Holy and Righteous One and asked that a murderer be released to you. You killed the author of life.

It is ironic that many Muslim societies became intolerant with the passage of time while Western nations became more accepting of the tolerance and pluralism of Islam. Islam accepts as worthy of salvation all those who believe in one god as the Master and Creator. In Chris-tianity, Jesus is the only route to salvation: "For God so loved the world that he gave his one and only Son, that whoever believes in him shall not perish but have eternal life. Whoever believes in him is not condemned, but whoever does not believe stands condemned already because he has not believed in the name of God's one and only Son."

In contrast to other great religions' attitudes toward nonadherents, Muslims accept Jews and Christians as "people of the Book." Thus Muslim global terrorists, including Osama bin Laden, display a strik-ing ignorance of Islam. They distort the message of Islam while at the same time using the name of religion to attract people to a path to terrorism. Bin Laden claims, "The enmity between us and the Jews goes far back in time and is deep rooted. There is no question that war between the two of us is inevitable." This comment contradicts 1,300 years of peaceful coexistence between Muslims and Jews, specifically in the Middle East and Spain. In fact, relations between the two communities were historically quite good. Indeed, when the Jews of Spain were expelled during the Inquisition, those who fled chose—almost without exception—to relocate in Muslim nations, where they knew they would be welcomed and accepted, and actually were.

The same sort of bigotry, inconsistent with the teachings of the Prophet and the tenets of Islam, was recently presented by Anjum Chaudri, a radical British mullah, in a BBC interview with Stephen Sackur. Chaudri rather remarkably said, "When we say 'innocent people' we mean Muslims, as far as non-Muslims are concerned . . . they have not accepted Islam, as far as we are concerned that is a crime against God." He went on to say that "you must hate and love for the sake of Allah. . . . I must have hatred toward everything which is not Islam." His statements are a contradiction of the Islamic message, which considers believers in one God to be "Muslims" and accepts the sanctity of all the primary religious texts of Judaism, Christianity, and Islam. Nevertheless, these hateful misinterpretations receive media attention and thus become part of the infectious distortion of Islam.

Recently on American television, the right-wing commentator Ann Coulter created a great stir by suggesting that Jews need to be "perfected" and by being perfected would become Christians. She repeatedly called Christians "perfected Jews." There is no parallel concept of exclusion anywhere in Islamic holy texts and doctrine. In Islam, all monotheistic religions are seen as paths to salvation. In Islam, Muslims, Jews, Christians, and all those who believe in a monotheistic god will be judged by their human conduct while on earth by God and not on the basis of the specific religion that they practice.

·

I was brought up in a home of gender equality. My parents had the same expectations for their two daughters as they did for their two sons. We had the same responsibilities and the same obligations. Most important, we had the same opportunities. All four of us—two sons, two daughters—were given the best education possible and were not only permitted but expected to attend the greatest universities in the world and to study equally hard. All four of us were brought up to believe that we had an obligation to repay society for all that had been provided to us by serving our people, by giving back to our society. We were encouraged to have careers, and specifically careers that benefited people. We were taught never to accept artificial limits on what we could be in life.

Equality for girl and boy children was certainly the norm in my family. My grandfather Sir Shah Nawaz Bhutto and my father, Zulfikar Ali Bhutto, were adamant in their decision to educate their daughters, reversing centuries of Sindhi (and other Muslim community) tradition. Among Bhuttos, sons and daughters were equal.

And when the time came to pick up my father's mantle and legacy and lead the Pakistan Peoples Party, I, as his eldest *child* present in Pakistan, led the struggle for democracy. No one among my father's followers opposed this on the ground of gender. This was the gender equality in Islam under which I was brought up. It is the gender equality that has been passed on to my son and two daughters. And I know it to be the gender equality that is specifically provided for and endorsed by Islam.

Understanding of the role of women in Islam requires that the Muslim Holy Text be looked at in the proper context. The Quran was produced at a time in history when women were seen as unequal in almost every society, especially in the Arabian Peninsula, and were often considered to be nothing more than slaves. Girl newborns were sometimes buried alive. We were taught that Islam stood for the liberation of humanity from the age of idol worship and darkness. Islam prohibited the killing of girls and gave women the right to divorce, child custody, alimony, and inheritance long before Western societies adopted these principles. Islam stressed the importance of education and knowledge, of compassion and taking care of the weak, the poor, and the dispossessed. Thus the message of Islam is pro–women's rights.

But with the passage of time, tribal values reasserted themselves. Social conditions in the Middle East, Africa, and Asia during medieval times deteriorated, causing much of the gender equality regression in Islamic societies that we see present today. Practices such as the wearing of the burqa, the isolation of women in their homes, female circumcision, and the banning of girls' education all come from various tribal traditions that have no basis in Islam. The traditionalists on gender issues, in fact, became social revisionists, defining women by men (husbands and fathers) and using deliberately narrow interpretations of law to exclude women from the consultation required by the Quran. In other words, the political effect of gender discrimination has been not only to socially discriminate against women and girls but also to limit the pluralistic requirements of the Holy Book.

Thus, "by the time the *shari'at* began to be codified, all sorts of pre-Islamic and non-Islamic influences, e.g., non-Arab societies like the Hellenic and Sassanid civilizations, had affected the thinking of Muslim jurists." (The Sharia is Islamic law and is composed of interpretations of the Quran and lessons learned from the Prophet's life [Sunnah]. These are continually made and modified, and thus the Sharia is dynamic and continues to change.)

So we must interpret the message of the Quran according to our given situation. We must look at the principles to see how they will guide us in practice as society evolves and new challenges arise. Quranic passages are in almost all cases intellectually provocative. Muslims are exhorted to apply reason, to consult, and to reach consensus on the application of quranic principles that are in almost all cases general, broad, and flexible, inviting discussion, interpretation, and flexibility.

The Holy Book establishes the common origin of man and woman, making them equal under God. Men and women came from the same being (*nafs*). The Quran:

O people! be careful of (your duty to) your Lord, Who created you from a single being and created its mate of the same (kind) and spread from these two, many men and women; and be careful of (your duty to) Allah, by Whom you demand one of another (your rights), and (to) the ties of relationship; surely Allah ever watches over you.

Since men and women came from the same being, neither gender has an inherent superiority over the other. They are equal. All the children of Adam (the common ancestor of all humans) are honored, regardless of sex. The Quran says: "And surely We have honored the children of Adam." Because both are honored equally, they must be equal in God's eyes and in Islam's practice.

With respect to the rights of husbands and wives in Islam, Maulana Azad, a well-known South Asian scholar and political leader, argues that husbands and wives are equal in terms of rights. The Quran: "They have rights [in regards to their husbands] similar to those against them in a just manner."

Sadly, this ideal is hardly the standard in all Muslim communities in the world today. Some insist on continuing to treat women as objects, subjugating women as they subjugate pluralism to a revisionist-traditionalist dogma that is more political than religious. One recent example is the highest Muslim cleric in Australia, Sheikh Taj el-Din al-Hilali. He gave a shocking Friday sermon in 2006 in which he (unbelievably) compared uncovered women to meat left out for a cat to eat: "If you take out uncovered meat and place it outside . . . and the cats come and eat it . . . whose fault is it, the cats' or the uncovered meat?" As we have seen, there is no support for al-Hilali's view in the Quran.

Al-Hilali argued that women who are not veiled and who are sexually assaulted are to blame for the attack. This inflammatory argument does not represent Islam in any way. Women are considered to have equal rights in the Quran and must be respected. But the same extremists, who close schools for girls, justifying their action in contradiction to the Quran, also rationalize violence and terrorism against innocent civilians, also in clear contradiction to explicit quranic doctrine. And not coincidentally, these are the same extremists who decry "democracy" as a foreign import inconsistent with Islam. Actually, it is sexual discrimination, terrorism, and dictatorship that are incongruent with the teachings of the Holy Prophet and the words of Allah.

Islamic precepts require women to dress modestly but do not ask that we cover up in any specific manner. Claiming that by not covering up, women are culpable for whatever befalls them is a perversion of logic. In fact the veil or the burqa, the all-enveloping chador, have more to do with tribal traditions. In the traditional past affluent women never left their homes other than to attend the weddings or funerals of relatives, and when they did they would cover themselves from head to foot. This was a norm of the time and the area but is not a precept of Islam in any way. There is nothing in the Quran that would suggest that this was the place of women at the time of the Prophet. Indeed, everything we know historically and everything that references the Prophet's family suggest a social system directly contrary to the traditionalists' view of women.

With respect to directives related to women's dress, there are two main verses that are often quoted to support the covering of women. These are the ones that jurists use when demanding that women wear the *hijab* (head scarf). When the Prophet went to Medina and the first Islamic state was established, the inhabitants of the city would come to his house at all times. The following passage in the Quran was revealed at that time:

> *O Prophet! say to your wives and your daughters and the women of the believers that they let down upon them their overgarments; this will be more proper, that they may be known, and thus they will not be given trouble; and Allah is Forgiving, Merciful.*

This example from the Quran is meant specifically for the family of the Prophet. In contrast, passages of the Quran meant for all women speak to all women without restriction.

When I was growing up, people in parts of my country were so poor that they could not afford clothes. Little children ran around naked in the villages. Often adults had a small cloth piece tied around their body. Poverty is still rampant, and often people cannot afford clothes or shoes, and walk with bare backs in parts of the world. One can imagine how much less cloth, or the ability to buy cloth, was present fifteen hundred years ago. It is in these social conditions that I read the following verse: "Say to the believing men that they cast down their looks and guard their private parts; that is purer for them; surely Allah is aware of what they do. And say to the believing women that they should lower their gaze and guard their modesty." This verse is clearly gender equal, not discriminating between men and women. The passage does call for modest dress, but for both sexes.

When I lived in our village in Larkana as a child in the late 1950s and early 1960s, the women in my family never left our home. The shopkeepers would come to us, as would the rest of the world. When I reached the age of puberty, my mother asked me to wear a burqa. Suddenly the world looked gray. I felt hot and uncomfortable breathing under the confines of the cloth. My father took one look at me and said, "My daughter does not have to wear the veil." My mother decided that if I was not to wear the all-enveloping burqa, she too would not wear one. So a departure from tradition took place.

There is a famous saying of the Holy Prophet that "the best veil is the veil in the eyes." That means that men should be God-fearing and look at women with respect.

Women's equality in Islam is not only in terms of political and social rights, but in religious rights as well. The Quran states:

Surely the men who submit and the women who submit, and the believing men and the believing women, and the obeying men and the obeying women, and the truthful men and the truthful women, and the patient men and the patient women and the humble men and the humble women, and the almsgiving men and the almsgiving women, and the fasting men and the fasting women, and the men who guard their pri-

vate parts and the women who guard, and the men who remember Allah much and the women who remember—Allah has prepared for them forgiveness and a mighty reward.

Women can have the same positive religious qualities as men. The Quran emphasizes this point by giving ten examples. This was certainly not the case in every religion, especially at the time the Quran was revealed, but in Islam there is no distinction between the sexes. In fact, it's quite remarkable, given the social conditions of seventh-century Arabia, that Islam advocated religious gender equality that took the rest of the world centuries to reach. However, sadly for me, the message of emancipation of women proclaimed by Islam is hardly to be seen in most Muslim countries. Instead tribal traditions and the values of subjugation spawned by authoritarian systems have robbed women of their Islamic rights to gender equality. One of the most essential components of religious pilgrimage known as Haj and Umrah is the Tawaf, the circling of the Kabah seven times. And from the birth of Islam to this very day, the millions who circle the Kabah are men and women together, together as equals under Allah. There is no separation; there is no hierarchy. There is equality.

The Quran goes further, giving women property rights: "men shall have the benefit of what they earn and women shall have the benefit of what they earn." Women can earn money, and what they earn is theirs. This gives them an autonomy almost unheard of in many parts of the world in the twenty-first century. And the Holy Book, the words of Allah revealed to the Prophet, consistently goes out of its way to make the point of sexual equality. Stylistically and grammatically, it would have been sufficient for the Quran to state that "all" or "all people" will be allotted what they earn. But it underscores the deliberate and emphatic gender equality by repeating the prescription for men and for women, making it clear, so there can be no confusion, that not only are men and women equal but women are clearly demarked as wage earners. In Islam, a women's place is not necessarily always in the home. It was not true for the Prophet's wife, who was a successful businesswoman, and it has not been true for me, as I have worked from the day I left university.

·

In addition to dress codes, there are Islamic practices concerning the conduct of women. In some Islamic countries, such as Saudi Arabia, women are not allowed to go outside without a male guardian, but this prohibition is not quranic. It came from later Muslim jurists. This is another example of Muslim jurists prescribing activities or restrictions based on their interpretations of the Quran, which are influenced by the society and culture in which they live and not shared by other jurists. Thus, even though the Quran does not put such a restriction of movement on women, some jurists have added these restrictions due to the context in which they were living. The variety of interpretation in Islam led to the establishment of four main schools of interpretation of Muslim law: Hanafi, Shafi'i, Maliki, and Hanbali.

Contemporary Muslim jurists can also apply this mind-set to contemporary society. However, somehow over the passage of years the dynamism of Muslim societies that permitted free, rational discourse to thrive has suffocated through the advent of tyranny. Over and over again in the Quran, Muslims are instructed by Allah to think, in more points of the Holy Book than we can count. Yet a long litany of Muslim rulers has made it, and in some cases continues to make it, impermissible to do anything but follow that which the ruler dictates.

The fall of Baghdad, which marked the end of the Abbassid Empire in the thirteenth century, also marked the end of the Golden Age of Islam. During the Golden Age of Islam, mathematics, philosophy, science, and Islamic theology were all studied at the highest levels in the world. Intellectual thought flourished, and pluralism in both thought and the makeup of society was encouraged. The fall of the Abbassid Empire to the Mongols brought in a period of tyranny from which the Muslim world has yet to escape. This political tyranny was accompanied by religious tyranny. The gates of *ijtihad* (interpretation of the Quran) were closed, and emphasis was put on memorizing the Quran and studying what interpretations already existed. It was this suppression of tolerance, pluralism, knowledge, reason, and the right to debate, discuss, and evolve a consensus that led to the decline of the Muslim power and glory witnessed at the zenith of the Islamic renaissance.

A free and *ijtihad*-oriented theology gave way to dogma, usually backed by the power of a court and army. In some ways, the stifling of the democratic spirit in theology is what started the beginning of the end of Islam's golden years. In Andalusia, Muslims prayed and fasted, but they also wrote and invented. The great library of Baghdad, sacked by the Mongol hordes in 1258, was said to have more than one million manuscripts on all disciplines of that age, from philosophy to medicine, history, and physics. The Ottoman and Mughal empires continued with conquest and were extremely rich, but they did not encourage expansion of knowledge. When printing was invented in 1455, the Ottoman Sultan Bayazid II (1481–1512) refused to allow it to be introduced into his empire. The Mughal emperor followed suit. Printing was not allowed in Muslim lands until 1727, setting the Muslim world far behind in knowledge as learning and debate expanded in Europe.

It is ironic that as Muslim societies were giving up their world intellectual leadership, the values of debate and discussion were increasingly being adopted by Western societies. It was an intellectual inversion, a reversal of the past. In medieval Europe, medicinal or scientific discussion was often considered heresy and not permitted even as great Muslim scientists and intellectuals were exploring new ideas. For example, Mohammad ibn Musa al-Khwarizmi, a Muslim mathematician who worked in ninth-century Baghdad, is considered one of the fathers of algebra, and in fact the word itself is derived from the title of one of al-Khwarizmi's books.

There is a good lesson in the Quran regarding fresh ways of thinking. Often in the Quran, men who reject God's message do so because their forefathers did not think this way. However, the Quran criticizes this type of thinking, emphasizing that one should not reject something out of hand just because it is different. Asghar Ali Engineer, a contemporary Muslim theologian, agrees and believes that we should also "escape from this rut" and begin to think afresh in our own experiential context and in the light of the values and principles laid down by the Quran.

It is clear that the path to growth and achievement lies in the values of tolerance, reason, and pluralism, values that are at the heart of Islam. The Quran centers on the quest for knowledge. There is nothing in the Holy Book that would condone ignorance.

.

The Holy Book specifically delineates the rights and protections of the weak, such as widows and orphans who come from poor families or whose property has been taken over by force by male relatives. The property of orphans who were not adults was to be placed in the guardianship of family elders. The Quran warns against misappropriation of any property, especially the property of orphans.

Men who marry propertied female orphans are also warned against marrying them for their money and then siphoning off their properties. The Quran makes it clear that the property, even after marriage, belongs to the woman and must not be taken over by the husband on the grounds of marital bonds:

> *And they ask you a decision about women. Say: Allah makes known to you His decision concerning them, and that which is recited to you in the Book concerning female orphans whom you do not give what is appointed for them while you desire to marry them, and concerning the weak among children, and that you should deal toward orphans with equity; and whatever good you do, Allah surely knows it.*

In a marriage, a husband must compensate his wife for bearing his child and feeding it. In other words, child rearing and household tasks are not a woman's natural responsibility. They are a joint responsibility. If a woman exceeds the routine, and in some ways does more than her share, she is to be compensated by the man for working beyond her responsibility and for covering her husband's responsibility. Modern society in the West now makes allowances for pregnant mothers who work. Unfortunately, this is rarely done in the Muslim world, which is where the recognition of a pregnant woman's right to compensation began. Also, in modern Western society governments are increasingly passing legislation guaranteeing parental leave to both mothers and fathers for a period of time after the birth of their child. Early child rearing is therefore recognized as a joint responsibility of both mothers and fathers. It is time for this issue, and so many others that relate to a woman's place in modern society, to be revisited. "And if they are pregnant, spend on them until they lay down their burden; then if they suckle for you, give them their recompense and enjoin one another among you to do good."

Many of the extremist interpretations of Islam used to justify discrimination against women in modern Islamic society are based on the tribal notion, repudiated in the Quran, that men are somehow superior to women. In Islam a man is responsible for financially supporting his wife and children irrespective of whether the mother has an income of her own. This reflects the context of the time of the revelation of the Prophet, in which men were generally the principal wage earners in families. Clearly this was the state of society in Mecca when the Prophet was born in the sixth century. Women, however, certainly did earn money, most directly illustrated by the Prophet's wife herself. Applied to today's world, where women often earn equal pay for equal work, equality would seem to be the case in all financial matters.

This is one of the fundamental reasons that as prime minister I encouraged both the public and private sectors to help women get better job opportunities. I also created Women's Development Banks to fund women entrepreneurs, to make loans to women-run businesses, consistent with the Quran's support for women's achieving full rights, including economic independence.

·

A review of how and why different sects emerged in early Islamic society helps to put the condition of Islam in the modern era into historical context. Some of these divisions, now being painfully manifested in the sectarian civil war raging in Iraq, have divided Muslims from Muslims for more than 1,300 years.

Following the death of Prophet Mohammad, the Muslim community was ruled by a succession of leaders called caliphs. Shias believe that the first caliph was Ali. He was the cousin of the Prophet and married to the Prophet's daughter Fatima. Ali was the father of the Prophet's only grandsons. But Sunnis support Abu Bakr as the first caliph.

Sunnis and Shias are the major sects in Islam, and it is important to understand their theological and historical background. The difference between the Shias and the Sunnis arose mainly from the disagreement over who should follow the Holy Prophet as the leader of the Muslim community he had established. The Shias claim that the Prophet Mohammad, when returning home from his last pilgrimage, stopped and told his companions that he was appointing Ali, his cousin who was married to the Prophet's daughter Fatima, to succeed him. However, Sunnis believe that on his deathbed, the Prophet selected Abu Bakr, the father of one of his wives, to be his successor. The Shias say that while Ali was burying the Prophet, Umar (who would become the second caliph) called the Prophet's companions and elected Abu Bakr. Sunni Muslims believe that Abu Bakr was the correctly appointed caliph. Shiites believe that Ali was both the first caliph and imam.

Abu Bakr was followed by Umar and Uthman as caliphs. Upon Uthman's assassination Ali became the fourth caliph and was universally accepted as such, except by those who inhabited the land that is now modern-day Syria. In 657 Uthman's nephew, Muawiya, challenged Ali and his followers in battle for power. Though the battle was inconclusive, Ali's ruling administration soon started falling apart. Ali was assassinated in A.D. 661, and his son Hasan, who was also the Prophet's grandson, took over. He was convinced to abdicate by Muawiya, who became leader of the Muslims. Upon Muawiya's death his son Yazid attempted to take over, but those in Iraq favored Imam Hussain (the second son of Ali). Hussain went to Kufa, in Iraq, to mobilize support before proceeding to Damascus, the capital of the empire, to claim the caliphate. He was ambushed and killed in battle by Yazid's forces in Karbala. Yazid's forces denied water to Imam Hussain's family, including the women and children. They were brutally slain, and the Prophet's grandson was beheaded. His head was taken to the court of Yazid accompanied by his sister Zainab, one of the survivors of the massacre at Karbala.

Shiites see this as a seminal moment of their history. They see Ali and Hussain as the living symbols of the Prophet's family who were murdered. This marked the beginning of the Shia as a sect with a strong sense of persecution, tragedy, and martyrdom.

All Muslims believe that there are Five Pillars of Islam, which are the fundamental principles that make up the most basic requirements for life as a Muslim:

1. *Shahada* ("witness"): This is the declaration that all Muslims must make: "I testify that there is no god but one God, and that Mohammad is the messenger of Allah."
2. *Salat* ("prayers"): All Muslims must pray five times daily, facing Mecca.
3. *Zakat* ("almsgiving"): Muslims must give a certain percentage of their yearly income to the poor and needy.
4. *Sawm* ("fasting"): During the holy month of Ramadan, all Muslims must fast every day from sunrise to sunset.
5. *Hajj* ("pilgrimage"): A pilgrimage to Mecca, the location of the

holiest place in Islam, must be performed by every Muslim, if possible, once in his or her lifetime.

In Sunni Islam there are four schools of Islamic Sunni law: Hanafi, Shafi'i, Maliki, and Hanbali. One of the emergent sects within Sunni Islam is the Wahhabi sect, which is strong and predominant in the Kingdom of Saudi Arabia today. It was founded by the eighteenth-century philosopher and cleric Mohammad ibn Abd al-Wahhab (1703–1792), a follower of the Hanbali school of Islamic law.

Wahhab collaborated with Mohammad ibn Saud, an Arab chief, and together they conquered the Arabian Peninsula. Although this area became part of the Ottoman Empire, their families later regained control of the territory and founded the modern state of Saudi Arabia. The House of Saud constitutes the Saudi monarchy, and Wahhabism is the official Islamic practice in the kingdom.

The Wahhabis destroyed many cemeteries of non-Wahhabi Muslims considered saintly and at whose graves others would go to pray. They would not allow people to pray at the graves of imams or saints. They made attendance at public prayer mandatory. In particular, Wahhabis were intolerant of the Shias. In 1802 Wahhabi forces sacked Karbala, in modern-day Iraq, and massacred every identifiable Shia man, woman, and child. Karbala is one of the holiest cities in Shia Islam, the site of the massacre of the Prophet's grandson Imam Hussain and other members of the Prophet's family.

Wahhabism is an austere and strict sect. It rejects Christians, Jews, and even some Islamic sects as "apostates." Some Wahhabis claim that it is a religious duty to kill Shias, leading to civil war and bloodshed in the heart of Islam. For most Sunnis "the most serious reason for opposition, however, was the Wahhabis' violent means of attempting to enforce their system and spread it to Muslims everywhere."

The Deoband school is a Sunni sect founded in India in the nineteenth century, named after the town in India where it was founded during the days of British rule. The Deoband is a conservative seminary that still functions. It gained acceptance after the unsuccessful 1857 Sepoy mutiny against British rule in India. Deobandis worry about Western influence on Muslim values and argue that instead of becoming "Westerners," Muslims must keep their own identity. They follow the Hanafi school of Islamic jurisprudence and look to discard non-Islamic practices.

The Deobandis believe that there is no distinction between religious and worldly affairs, that Islam is part of every aspect of life. Increasingly, the Wahhabi movement from Saudi Arabia has invested in Deobandi schools to gain ingress into South Asia.

But the main funnel for Wahhabi funds in the modern era has been the Jamaat-i-Islami political and social movement, founded by Maulana Maudoodi. Most Wahhabi funding still goes into schools sponsored by the Jamaat-i-Islami. When I was a child, it was common to hear stories that money would be given to Jamaat-i-Islami indirectly. The Saudi clergy would buy up thousands of copies of books by Maulana Maudoodi, make a payment for them, and then dump them into the sea because too few people wanted to read them. However, this would change with the seizure of power through a coup in Pakistan by General Zia-ul-Haq, the military dictator who stole power on July 5, 1977.

The other major sect of Islam, the Shia, makes up between 10 and 20 percent of Muslims, depending on whom one talks to. They are located mostly in Iran, Iraq, Syria, Lebanon, Pakistan, the Gulf states, and India.

There are some significant differences between the Shia doctrine and Sunni Islam. The Shias believe in both temporal and spiritual leadership on earth. They believe that the imam is the spiritual head of the Muslim community, which also needs a temporal head. As the scholar Solomon Nigosian succinctly puts it, "All faithful Shiites believe that he [the imam] is, through his relationship to Muhammad, the divinely appointed ruler and teacher who has succeeded to the prerogatives of the Prophet himself. Moreover, the imam possesses superhuman qualities, more particularly a 'divine light' [i.e., superhuman knowledge], which is transferred to him from Adam through Muhammad and Ali."

Shiites observe Ashura (Arabic for "ten"), to mourn the assassination of Hussain and others in the Battle of Karbala. During this period they wear signs of mourning and recall the story of Imam Hussain setting off to Karbala with his small caravan of largely family members. They mourn the murders that took place in the run-up to the murder of Imam Hussain and, of course, that of Imam Hussain himself.

The Karbala tragedy is, for Shia Muslims, the lesson history teaches of the price that must be paid for following the path of truth and for resisting tyranny. In every generation, it is said, there is Karbala when people rise up against a powerful tyrant, knowing they are outnumbered but unable to remain silent in the face of opposition.

Shias offer Fataha, the Muslim prayer, at the burial places of imams and their descendants. Many Shiites go on pilgrimages to the burial places of important imams to pray to God that in the name of the beloved Imam Hussain they may be blessed to overcome their troubles.

There are three main Shia groups: Imami, Zaydi, and Ismaili. Imami, the largest of Shia sects, is also known as the Twelvers. The Imami do not recognize the first three caliphs that Sunnis recognize (Abu Bakr, Umar, and Uthman). They believe Ali was the first imam and after Ali the imams continued in the Ali-Fatima line (Fatima was the only surviving daughter of the Prophet, and Ali was her husband and the father of the Prophet's grandchildren). Imamis believe that the twelfth imam was Mohammad al-Mahdi, and consider him to be the "Hidden Imam." Twelvers consider al-Mahdi to be hidden because he did not die but went into occultation and is still guiding human affairs. They also believe that he will return someday.

Imamis believe in human free will, since God sent imams to guide people and thus people can freely choose the correct path. Imams are members of the Prophet's family, so humans cannot make a mistake in selecting the imam. In Shia Imami belief, the return of the twelfth imam to earth will bring about the success of Islam and the ascendancy of monotheism, and will serve as God's final judgment, congruent with the return of Jesus to Christians or the coming of the Messiah to Jews.

While there are theological differences between the Shias and the Sunnis, there is also the very real possibility for constructive coexistence. I, like many other Muslims, had a Sunni father and a Shia mother. When we were growing up, Sunnis and Shias lived in peace with each other. The Muslim month of Muharram was observed by both Sunnis and Shias. However, in the 1980s, during the dictatorship of General Zia, a sinister campaign began to divide the Muslims in Pakistan, and indeed Muslims everywhere, by pitting Shia and Sunnis against one another.

I remember my days as a political prisoner in Karachi Central Jail in 1981. A prison warden, bringing me my food, said she was frightened. She had heard that a Shia family had moved into the lane of the house where she lived. "You know Shias eat children," she said. "I lock my child up all day so that Shias can't kidnap, kill, and eat my child." She had heard this from the imam at her local mosque. How interesting that it parallels the "blood libel," so pervasively accepted in parts of Europe through history, that Jews drank the blood of Christian children. I never cease to be amazed by the horrific mythologies that are spread in the name of God.

Pernicious propaganda has been spread against Shias for generations. In 1979 there was a successful revolution in Iran led by the Shia religious leader Ayatollah Ruhollah Khomeini. There was fear that Shias could rise up in other Muslim countries. To stop the spread of the Iranian Revolution, a plan was apparently made by Zia's forces to demonize Shias and to pit Shia against Sunni. Thus we can see the secular political motivation for encouraging conflict between theological branches of Islam. (It would certainly also seem to be the case in contemporary Iraq's sectarian civil war.)

In Pakistan's largely Shia northern region, the extreme Sunni Wahhabis had begun to come from the Arab states to join the Afghan mujahideen fighting the Soviets in nearby Afghanistan. Coincidentally, Herat, close to the southern Pakistan border, where the resistance was being organized, was also Shia. Anti-Shia sentiment was encouraged to motivate mujahideen to fight. An uprising by the Shias in northern Pakistan was brutally put down by the army in what many northerners today describe as genocide. The Shias reacted and began retaliatory killings against Sunnis. The ugly face of sectarian division had raised itself in Pakistan. It later spread to Iraq, where it had been ignited by Shia-Sunni passions that flared during the battle between Shia Iran and Sunni Iraq in the 1980s.

The second major subgroup of the Shia is the Zaydi, whose name comes from the fourth imam, Zayd ibn Ali (d. 740). The group used to have many adherents, but it is now limited to a small sect of followers located mostly in the Yemeni highlands. They do not view the first three caliphs as usurpers but think imams can be only from Ali's line. They do not see imams as infallible and in fact believe they can be deposed. They do not believe in the Hidden Imam. They believe that the imam is someone who is guided by God but that the imam does not possess a "divine light," which is a belief held by the other Shia sects. Their prayer and ablution are different from those of other Shiites. They don't allow mixed marriages outside of their own Shia subsect, and they are not allowed to eat meat slaughtered by non-Muslims.

The third main sect of Shia Islam is the Ismaili sect, also called Seveners. There is agreement about the sixth imam, Jafar al-Sadiq, who died in 765. But Ismailis do not recognize the Musa as the seventh imam, as Imamis do. Instead they believe that Ismail was the seventh imam. They believe that Ismail's son Mohammad was the next imam after Ismail. Some Ismailis believe he disappeared and will return (similar to the Imami belief that Mohammad al-Mahdi will return). Others believe there is still a line of imams of which the Aga Khan is the latest. (I know that all of this must seem like relatively insignificant divisions, but I believe we'd find the same level of variance between groups within Christianity and Judaism as well.)

For my generation in Pakistan, the differences between the sects of Islam seemed insignificant. We were brought up to believe that all Muslims believe in one Islam, face in one direction to pray, recite one Quran, and follow the Prophet Mohammad as the last Prophet. Tolerance within our own religion, and with other religions, was the touchstone of our belief.

This unity of the Muslim ethos was broken in Pakistan under General Zia. Suddenly Muslims were asked to indicate whether they were Sunnis or Shias on government-issued documents. General Zia also decided to end the separation between mosque and state by promulgating a slew of religious laws and a religious tax. Pakistani Muslims were required to declare which sect they belonged to as these laws were applied. Suddenly, instead of being Muslims who believed in one Allah, one direction of prayer, one Quran, and one last Prophet, we had to identify ourselves as Sunni Muslims or Shia Muslims.

Then the Sunni majority began to discriminate against the Shia minority. What General Zia began in Pakistan when training the Afghan mujahideen—exaggerating internal differences between Muslims for political reasons—has now, like a cancer in the body, spread to Iraq. Abu Musab al-Zarqawi, the well-known Iraqi insurgent, was one of the recruits of the Pakistani Directorate for Inter-Services Intelligence (ISI) during the Zia years.

It is my firm belief that until Muslims revert to the traditional interpretation of Islam—in which "you shall have your religion, and I shall have mine" is respected and adhered to—the factional strife within Muslim countries will continue. Indeed, until quranic tolerance is reestablished, the key Muslim countries of Pakistan and Iraq will not only continue to weaken them but will continue to threaten to spread inflexible and extremist interpretations elsewhere in the Muslim world. Those who teach the killing of adherents of other sects or religions are damaging Muslim societies as well as threatening non-Muslim societies.

Most modern Muslims accept that sectarian differences do not warrant distrust and violence between sects. The sectarian divide that has lasted with varying degrees of intensity for 1,300 years must now be bridged to get beyond violent expression and its terrible consequences. The intensity of distrust and discrimination seems, at least in part, to be correlated with political events. The Iranian Shia Revolution of 1979, the Zia military dictatorship in Pakistan using extreme Sunni interpretations for political ends, and the Iraq-Iran War of the 1980s are key historical events cast as "Sunni versus Shia" sectarianism. They have become the foundation of a perceived threat that continues to breed fear and intolerance.

·

Another of the misconceptions about Islam that many in the West hold—encouraged by the images of extremists and militants who reject modernity—is that Islam is incongruent and incompatible with science and technology. I believe that a careful reading of the Quran proves just the opposite.

One of the critical elements of Islam is the importance of knowledge for healthy societies and for civilization. The Quran instructs its adherents to seek knowledge. The Holy Book does not ask Muslims to just memorize the Quran but instead wants humans (God's creation) to continue to interpret the world and to continue to seek information:

> The Ummah {Muslim community} was built on the foundation of tawhid, istikhlaf, *the pursuit of knowledge, and personal and communal responsibility. Although it {the Ummah} was once a leading creator of and contributor to human civilization, over the last few centuries it has become weak and backward to the point of crisis.*

Many people may ask why the Ummah has not always dealt successfully with modern concepts of science and technology. Why hasn't the community more broadly risen to the challenges and opportunities of modern technology? Clearly, science and technology cannot simply be purchased from the West. To fulfill the extraordinary potential of the Muslim community, especially for the next generation of Muslims, knowledge and innovation in science and technology must be developed, understood, and embraced. The early Muslims created a great society that was not just a great religious community but also one of the world's most eminent civilizations. The Quran and its messages "transformed the early Arab Muslims into a great people, the bearers of a great religion, and the creators of great civilizations and history. They sought guidance in the Quran and developed their means through reason and knowledge of Allah's natural laws."

But following the centuries when Islam led the world in culture and innovation, in medicine and literature, while Europe was mired in the Dark Ages, the nations of Islam lost their competitive edge. Over time, far too many political and religious leaders became oppressive and inflexible. They resisted experimentation and change. The clerics felt they should have the power and authority to interpret the Holy Book. The rulers felt that free inquiry challenged their right to autocratic power. In any case, intellectuals were no longer allowed to pursue knowledge freely or to engage the practical world, and tragically they turned inward.

We can only imagine how very different the course of civilization would have been if Islamic scholarship, innovation, science, and medicine had been encouraged to develop and grow over the centuries. But the intellectual strangulation of the rulers and clerics stood in the way of creative, innovative thought. Scholars turned away from seeking new knowledge. Jurists and intellectuals focused on following the letter of the law in Islam as interpreted by a few chosen scholars, rather than reading the Quran and thinking for themselves. Instead of seeking meaning from the Quran to marry with reason and consultation to develop a new framework for contemporary challenges, the easy path was taken by recourse to meanings interpreted by scholars of a bygone age.

As Muslim children we are taught this famous story of the Prophet and his grandson Hussain: One Friday, the Prophet was leading prayers when his grandson Hussain climbed on his back. Rather than throw him off immediately so that he could keep praying in the prescribed manner, the Prophet allowed the young boy to play for a bit and then carefully put him down. He explained later that he did not want to rise hastily as this would not have allowed his grandson to enjoy the ride on his back. The Prophet fulfilled his obligatory prayer to God, but he also allowed his young grandson to explore and to play. The story demonstrates that adherence to Islam does not mean diverting attention from all else. God created humans with the capacity for thought and knowledge so they could use them, not go through life with blinders on. The Prophet remarked on the importance of seeking knowledge throughout life: "Seek knowledge by even going to China, for seeking knowledge is incumbent on every Muslim." The Prophet placed the utmost importance on seeking knowledge, instructing humans to go to extraordinary lengths to gain not just religious knowledge but all knowledge. Mehdi Golshani, a renowned contemporary Iranian philosopher and physicist, agrees, stressing, "The concept of knowledge was strong during the golden age of Islamic civilization. They got this concept of knowledge from the Quran and Sunnah. That's why they acquired knowledge from other nations as well and added to it."

All fields of knowledge are considered important, especially knowledge of the world, because knowledge of the world is knowledge of God's creation. In this way a Muslim, through science, knows and understands the signs of God that are everywhere. The Quran states: "And one of His signs is the creation of the heavens and the Earth and the diversity of your tongues and colors; most surely there are signs in this for the learned." This passage indicates God's desire that we learn about nature and thus engage in scientific endeavors.

This next verse continues: "Say: Travel in the Earth and see how He makes the first creation, then Allah creates the latter creation; surely Allah has power over all things." It is God's desire that humans explore their world. Essentially the Quran is arguing for the study of science. According to Golshani, science and religion have the same goal, which is to find God. In science one looks for God in his creation. By studying science, one is looking at God's creation. Even the earliest human being was a scientist: "And He taught Adam all the names, then presented them to the angels; then He said: Tell me the names of those if you are right." Adam is asked to comprehend the world, and thus Adam serves as the first example of how to examine our world through science.

Lastly, God created in humans intelligence and a sense of comprehension so that we could use it: "And Allah has brought you forth from the wombs of your mothers—you did not know anything—and He gave you hearing and sight and hearts that you may give thanks."

I know that there are some people who argue that technology and modernization will lead to increased secularization. Some in traditional societies resist modernization for that very reason. To them the inherent dangers of technology include the danger of secularization, the loss of an Islamic culture, and even Westernization.

But religion can use technology to help preserve the Muslim community. As commanded by the Five Pillars of Islam, Muslims pray five times a day. Traditionally, at prayer time, a *muezzin* (the man in charge of a mosque, also called imam) calls the faithful to prayer by reading out the rhythmic "call to prayer" from the minaret of a mosque to let Muslims know that it is prayer time. However, "in almost every Islamic community today, the loudspeaker, radio and television have become essential in the traditional call to prayer, a remarkable juxtaposition of high media technology and conservative religious practice."

In the 1960s in Singapore there was significant urbanization and resettlement. During this process many people were relocated from their traditional villages, which had been made up mostly of homogeneous religious groups. Once they were in the cities, the call for prayer from a mosque was not loud enough for people in the neighborhood to hear it, so the Muslim community installed loudspeakers so Muslims could hear the call to prayer. Thus a modern technology invented in the non-Muslim world was used to keep the Muslim community together, not separate and destroy it.

Once in this new urban space, Muslim communities were no longer in homogeneous groups. Many different sects and religious groups lived together, and some groups did not appreciate the call to prayer. They considered it noise pollution. An agreement was formed, through peaceful discussion, between the religious groups and the government to lower the volume on outward-facing loudspeakers on the mosques and to build all future mosques with inward-facing speakers. Last, and most important, the call to prayer began to be broadcast over the radio.

In the new pluralistic community it became difficult to maintain the acoustic Muslim community. Thus the Muslim community was able to take advantage of modern technology to achieve a closer community. Airwaves replaced the traditional imam's call to prayer and were used as a tool for communication among communities. The radio helped maintain the acoustical space traditionally occupied by the imam of the Mosque. The importance of the call to prayer for the Ummah cannot be overstated. Maintaining the call to prayer was important for keeping the community whole: "The call to prayer does not merely inform Muslims that it is time to pray, it is a statement that says 'we are Muslims.' "

Many have argued that the phenomena of mass media and technology alienate the individual from the community. But in the case of Muslim communities, radio broadcasts and indeed now television and the Internet are able to bring communities together, demonstrating how technology can be used to sustain a community's identity in an increasingly pluralistic and diverse society.

Another current technology that is being used to keep religious values strong in the Muslim community is the use of cell phones. Some modern cell phones have automatic alarm clock–type reminders for the call to prayer five times a day so that if the user is not near a mosque, he or she can still know when to pray. The cell phones also have an automatic silence mode that allows Muslims to avoid interruptions while praying; a Ramadan calendar with estimated fasting times; and a compass to indicate the direction of Mecca, so one can pray in the correct direction.

Certainly Muslims are openly encouraged to seek medical attention and treatment, take medication, get advanced surgery, eat nutritiously, and take advantage of modern medical technology. Even those who decry technology as a Western incursion on Islamic values accept modern medicine. Many people believe that the greatest extremist and the most infamous Western-hating terrorist on the planet, Osama bin Laden, stays alive (wherever he is) by kidney dialysis. So it would seem that the extremists who reject modern technology as a Western concept apply it selectively and conveniently.

·

"Neither Islam nor its culture is the major obstacle to political modernity, even if undemocratic leaders sometimes use Islam as their excuse."

It is interesting that we even have to address the question of Islam and democracy. I am not aware of any substantial intellectual inquiry into "Judaism and democracy" or "Christianity and democracy" or "Hinduism and democracy." Many people in the West look at the "Islamic world" and see very few democracies (see the discussion in chapter 3). Some assume this is a result of some inherent hostility to democratic ideals in the Islamic religion.

Let us first dispel a misconception that colors the entire discussion of Islam and democracy. The misconception is that Islam is a unitary, rigid social system transcending religion. Further, it colors Islam as an inflexible dogma that removes the element of choice from individual lives. In addition, the misconception suggests that there is little diversity in Islam. This view is persuasively challenged by scholars, historians, and reform clerics. Islam is a religion, not a unitary social structure. It is a religion that unites Muslim communities behind key theological and moral precepts. However, there is great diversity within the Muslim Ummah based on culture, history, tradition, and interpretation of religion. Islam's more than one billion adherents include many sects. The Muslim Sharia law is quite varied in size and content in the various Muslim communities and schools of thought around the world. Much as in other religions, one can take a small quotation from the Quran and then make larger judgments on "Islamic civilization's" ability to support democracy.

If the texts of other religions were judged this way, I would guess that none would be seen as open to democracy. For example, if we take the Christian Bible or the Jewish Torah and look at the principles or systems of governance discussed, neither would appear to be particularly democratic or pluralistic. As an example, stoning is a form of capital punishment that is a product of the Old Testament. Just as in Christianity, worship is only one part (albeit a very significant part) of a broader Muslim culture, a culture that can accommodate democracy. In early Muslim societies, the existing polity took cues and basic ideas from Islam and then applied them to their particular historical contexts, developing unique cultural and socioeconomic political systems. Modern Muslims can do the same within their societies. Just as modern Christians do not seek to re-create the Dark Ages, modern Muslims can use the Quran as a guide to modern life.

The past is used too frequently to define modern Muslims, especially when evaluating their receptivity to democracy. We don't define Judaism by the brutality of the conquest of the tribes of Canaan or by the pain and suffering of the plagues on Egypt. We don't define Christianity by the barbarism of the Dark Ages or by the persecution of the Inquisition. When analysts look at the receptivity of modern Muslim communities to democracy, they too often look to Islamic texts and interpretations, as well as to the kind of social structure of the first community of Muslims. This construct, labeled "Muslim exceptionalism," is based on the view that the norms of the Muslim community of the past must necessarily define the Muslim community of the present. It assumes that Muslim thought and Muslim society have not evolved, adapted, or changed over time. Some feel that "the character of Muslim societies has been determined by a specific and remote period in their past during which the social and political order that continues to guide them was established."

The scholar is referring to Prophet Mohammad's early community of Muslims in seventh-century Arabia. This theory is predicated on the bizarre belief that the strength of the past continues to hold on to the psyche of Muslim society, blocking progress in political and other fields, including human rights and technological and economic development.

The early Muslim caliphate (immediately after Prophet Mohammad's death) did not have Sharia law to govern it, as Sharia was not yet developed. The system was led by a rightly guided caliph selected by the people. The caliph would rule in the public interest and be accountable to the people. The Quran boldly proclaims that the rulers and the ruled were equal before the eyes of God. The ruler could demand no special privileges.

Much later, in medieval Islam, Sharia law was developed. The original Sharia law actually restricted autocratic rulers from abusing their people, through a system of checks to ensure justice and equality. Sharia forced the rulers to acknowledge a system of law that restrained their power, thus affording the people a minimal type of protection.

Some scholars argue that more than democracy and Islam merely being compatible, Islam as a religion contains more pluralism and justice than other religions and is in fact a more fertile ground for democracy. Abdulkarim Soroush is one of these scholars. He believes that "Islam and democracy are not only compatible, their association is inevitable. In a Muslim society, one without the other is not perfect." Soroush sees two main foundations for democracy in the Muslim world: freedom and evolution. He speaks of freedom of religion and freedom of thought. In order for one to be a true believer one must be free to select a faith and to leave that faith. The same concept of freedom is applied to a polity. The people (the majority) must be free to select their government and laws. While sacred texts are constant and divine, interpretations should evolve over time based on changes in the social and political environments.

Progressive scholars like Soroush argue that "Sharia is something expandable. You cannot imagine the extent of its flexibility." Clearly there are principles inherent in the Quran and intrinsic to Islam that can be applied through reason and consultation to reach a consensus about social and political conduct appropriate to changing times. Sharia is a way to get to justice; thus Sharia is not an end but rather a means to justice or equality or other principles that have been laid down.

The Quran provides broad beliefs and morals by which to live. The specifics were left to be interpreted in light of the proper historical context. "The text is silent. We have to hear its voice. In order to hear, we need presuppositions. In order to have presuppositions, we need the knowledge of the age. In order to have the knowledge of the age, we have to surrender to change."

Equally important to the context of interpretation of the Quran is who interprets it. Some Muslims, especially those belonging to theocratic regimes, try to assert that only a select few can interpret the Quran. This is not the case. Interpretation of the Quran is not limited to any one person or committee. The Quran did not establish a specific institution or group of leaders as its sole interpreters. Any Muslim is free to interpret the Quran. All Muslims are guaranteed the right to interpret the Quran (*ijtihad*). Thus even the approach to interpretation of the Quran is imbedded with democratic values. Indeed, Muslims are told that each person is accountable for his or her individual behavior. No relative, teacher, or other can intervene for a Muslim on the Day of Judgment.

The scholar Laith Kubba asserts that while historical quranic jurisprudence is important, "Islamic authority is the Quran's alone." He argues that "scripture has influenced these traditions, even formed them, if you like. But it has not *sanctified* them, and it is a mistake to think that it has done so." He sees interpretations—whether by government or other legal or quasi-legal bodies—as less important than the Holy Text itself.

Many of the Islamic traditions associated with Islam and public life are just that, traditions. Traditions are not the commandments of God as recorded in the Quran or the facts of the life of the Prophet of Islam, which is looked at as the life of the perfect Muslim. It is the principles that the Prophet lived by, which are everlasting. He lived simply and was kind, compassionate, and courageous. With respect to the role of women, although traditionally men and women have been separated in many Islamic societies, this separation is from traditional cultural sources that are outside Islam.

A similar pattern can be seen with respect to the question of government in Islam. Although Muslims traditionally look to the early community in Medina as an example of good Islamic governance, there is no directive in the Quran saying that cities have to be governed as Medina was governed. Nonetheless, Medina was governed by consensus, with the Prophet consulting with his companions on the matters of the state. Those principles of consultation and consensus were not only the democratic foundations of Muslim society at the time of the Prophet but also the principles that should guide Muslim political systems today. Unfortunately, the political and social structures of Medina at the time of the Prophet, which prevailed in and encouraged the innovation of the Golden Age of Islam, were gradually replaced with traditionalist, exceptionalism models that were resistant to adaptability and change. Many extremists today would like to freeze Islam in this rigid dogma, disregarding not only the pluralism of the Quran but the democratic governance of the Medina of the Prophet.

According to Abdulaziz Sachedina and many other scholars, the Quran is interpreted by humans and thus all interpretations and laws not explicitly spelled out in the Quran are not immutable. He asserts that "the Quran remains in the hands of humans who have to decide how to make it relevant to their moral-spiritual existence at a given time and place in history." The prominent Islamic scholar Khaled

Abou El Fadl agrees, citing this quranic reference: "And (as for) the man who steals and the woman who steals, cut off their hands as a punishment for what they have earned, an exemplary punishment from Allah; and Allah is Mighty, Wise." While this seems fairly un-ambiguous, according to El Fadl, people must interpret the passage before making law out of such verses. Muslims have to interpret the meanings of the words "cut off," "steals," and "hands." Even though the meaning might seem clear, how can one be sure that any human law is the same as the one intended by God? There is no way for humans to perfectly codify and carry out God's law and avoid any error. "Under this conception, no religious laws can or may be enforced by the state."

In my view, given that there were no jails at the time, the purpose of cutting off hands was not only to punish the culprit but to warn others in society that the person was a thief. Thus it was a punishment, a deterrent, and a security for the safety of others in the community; that is the principle of the verse. It is that metaphor of deterrence and warning that is valid to modern life, not the literal act.

Every interpretation needs to be based on the context in which it is undertaken. In the modern world, modern interpretations need to be made while respecting the underlying principles of the Quran. The Quran, while the word of God, is a text that is historically rooted in the time of its revelation. There is no explicit mention of democracy in the Quran because it was not a word in use in seventh-century Arabia. However, the principles of consultation and consensus among the people, which are found in the Quran, are the bases of democracy. Moreover, the principles of equality, justice, and law, which are the underlying foundations of democracy, are repeatedly stressed in the Quran.

The caliphs were "elected" by the community. Hazrat Ali did not challenge the "election" of Caliph Abu Bakr or the other caliphs. In time he too was "elected" caliph. If Ali's followers complained of the "election," it was on the grounds that it was not a fair election since one of the "candidates" was missing from the meeting. He would have liked to put forward his case, namely that he had been "nominated" by the Prophet as a "candidate," but it was still up to the community to select the caliph. The most important lesson of the emergence of the first caliph is that it was decided not by force or power but by the free choice of the Muslims present. And despite the reservations of those who supported Ali, he accepted the result of that election and subsequent ones.

For Muslims like me, who believe in democracy, Islam is about consent and people's participation. Islam and democracy are compatible. Radwan Masmoudi agrees that contemporary interpretations need to continue to be made; he asserts that it is better that "the doors of *ijtihad*—closed for some 500 years—be reopened." Even the conservative Pakistani Islamist leader Khurshid Ahmad concedes that "God has revealed only broad principles and has endowed man with the freedom to apply them in every age in the way suited to the spirit and conditions of that age. It is through the *ijtihad* that people of every age try to implement and apply divine guidance to the problems of their times."

Maulana Maudoodi, the leader of the Jamaat, was the spiritual father of the dictator Zia-ul-Haq. He had a close connection with the Saudi clergy. After the Soviets invaded Afghanistan, Zia turned to him for support in raising funds and recruiting fighters for the Afghan mujahideen. He turned to others across the Muslim world who shared his views on jihad, the West, and democracy. I grew up during Maulana Maudoodi's lifetime. He opposed my father's election. He was never taken seriously by the vast majority of the Muslims of Pakistan. The political arrangement camouflaged behind the veil of Islam and brokered by General Zia was complex and multifaceted. It was a dark coalition between Maudoodi, Pakistani intelligence known as the ISI (led then by Zia's relative General Akhtar Abdul Rehman), the Pakistani military, and the state, which would have far-reaching consequences in the politics of my country and the larger world.

Maulana Maudoodi dubbed the founder of Pakistan, Quaid-e-

Azam Mohammad Ali Jinnah, a *kafir* ("nonbeliever"). But the Muslims of India rejected Maudoodi and instead supported Mohammad Ali Jinnah and his more secular view of religion and politics. Maulana Maudoodi, viewing my father's politics as inconsistent with the extremists' agenda, also declared my father a *kafir* in 1970, but my father's party roundly defeated the Islamists in the election held that year.

When I ran for prime minister in 1988, Maudoodi's party called me a *kafir*, as they had my father before me. The people made their own decision on the attacks against me by the Jamaat when they voted overwhelmingly for the PPP and elected me prime minister, the first woman ever elected to head an Islamic state. It was a rejection of extremism and bigotry. Each major "wanted" terrorist following the September 11, 2001, attacks has been recovered from the house of a member of Maudoodi's party, including Ramzi Binalshibh, Abu Zubaydah, and Khalid Sheikh Mohammed.

Islam proclaims that the earth belongs to *"Khalq e Khuda,"* the people of God. We are all God's creatures. The earth is given to us in trust by God. We the people are the agents of God in this world. We are to govern the earth as a sacred trust and as trustees of that responsibility to pass it on to future generations. The right to declare who is a "good Muslim" and who is a "bad Muslim" is a right that belongs only to God. Those who say that we on earth must determine who is a good Muslim and who is a bad Muslim are in many ways responsible for the political legacy of murder, mayhem, sectarian warfare, and oppression of women and minorities we see in the Muslim world. These extremists are destroying the Muslim world by pitting Muslim against Muslim.

Yet before supporting General Zia, Maudoodi actually supported Fatima Jinnah, a woman candidate for president of Pakistan in the 1960s. So at the time he didn't find any quranic prohibition against women serving as leaders of Muslim societies. (His later antiwoman epiphany against my right to serve as prime minister was thus obviously more political than theocratic.) When Maudoodi endorsed

Fatima Jinnah, he stressed that the executive power in the government needs to be selected by the people. He saw the electoral process as integral and called for the principles below to be followed in selecting an executive. These principles are developed from readings of the Quran and Sunnah:

1. The election of the chief executive depends entirely on the will of the general public with no one having the right to impose himself forcibly as the ruler.
2. The ruler must be elected and can only rule by the consent of the people. No clan or class shall have a monopoly of rulership.
3. The election shall be free of all coercion.

These principles stress the important Islamic ideals of free and fair elections, as well as the power of the opposition and the will of the people. However, during the long night of terror when General Zia's dictatorship crushed all opposition, Maudoodi's party stood shoulder to shoulder with General Zia. Maudoodi's party leaders were members of Zia's cabinet. Thus we see once again that political expediency and not political principle was at the root of Jamaat's alliance with the brutal military dictatorship. Political power prevailed over Islamic principles of consensus and pluralism. It is part of a disquieting pattern all over the Muslim world that most often remains intact.

I've talked about a variety of authors who argue for the importance of focusing on the Quran's meaning for each age. Now I would like to present some quranic references to democratic ideals that can be used positively in current circumstances.

Importantly, the Quran gives authority to human representatives on earth (i.e., government): "O you who believe! obey Allah and obey the Messenger and those in authority from among you." This passage is the basic building block of government, as it allows for human authority on earth: government.

Shura (Arabic for "consultation") is an important quranic concept. The Quran instructs governors to consult the people whom they rule. This is the basis of consultative government in Islam. The Quran gives instruction to the Prophet to consult those whom he rules and to govern the people by a consultative, consensual process. This is good in God's eyes. According to the Quran:

> *Thus it is due to mercy from Allah that you deal with them gently, and had you been rough, hard hearted, they would certainly have dispersed from around you; pardon them therefore and ask pardon for them, and take counsel with them in the affair; so when you have decided, then place your trust in Allah; surely Allah loves those who trust.*

This verse explicitly says that God prefers consultation of the ruled people. It also instructs rulers not to be cruel. This next verse reinforces the idea of consultation in government: "And those who respond to their Lord and keep up prayer, and their rule is to take counsel among themselves, and who spend out of what We have given them."

The idea of *shura* did not simply mean seeking the opinion of elites in a society; "it signified, more broadly, resistance to autocracy, government by force, or oppression."

Additionally, democracy is supported in Islam by the concept of *ijma* (Arabic for "consensus"). The Prophet knew that the consensus of the people was superior to the rule of one man. The Prophet stressed this point by saying, according to the hadith, "My community will not agree on an error." This indicates God's preference for communal politics.

Democratic governance allows for a people to express opinions. Often, however, opinion is not singular but diverse. Part of democratic governance is not just to allow the majority to rule but also to allow space for a legitimate, formal opposition. Opposition ranges from violent rebellion and overthrow of government to disagreement within the ruling party. Clearly violent rebellion should not be accepted, but all types of peaceful opposition are acceptable means for an opposition to get its say, including providing a chance at ruling (through elections). A thriving and effective opposition is predicated on all parties within a system agreeing on the fundamentals, the rules of the game, the Constitution, and the methods of governance and regime change:

> *Modern democratic opposition involves a number of basic assumptions. In general terms, there is an assumption of a consensus among all in the system on the fundamental construction of that political system. If that consensus is absent, it is assumed that legitimate opposition will not lead to violent and military efforts to overthrow the existing system and that any opposition that comes to power will not so alter the system that it could not subsequently be restored by nonviolent means.*

Allowing some degree of political opposition has roots in the Quran. But while *fitnah* (disorder through schism or division) is not allowed, other types of disagreement are. By prohibiting division and disorder, the Quran establishes limits on what is allowed in opposition. By showing the limit the opposition can go to, the Quran provides a political space for the opposition. In further support of respecting oppositional points of view, the Quran mentions diversity (see below), revealing that God created, and thus desires, diversity: "And if your Lord had pleased He would certainly have made people a single nation, and they shall continue to differ." In other words, differences can and do exist and must be respected.

The Quran clearly commands the government to rule with justice:

> *Surely Allah commands you to make over trusts to their owners and that when you judge between people you judge with justice; surely Allah admonishes you with what is excellent; surely Allah is Seeing, Hearing.*

This next verse of the Quran instructs governments to rule with benevolence. It commands them to promote the public interest, to help the needy, and not to benefit the rich in society at the expense of the poor:

> *Whatever Allah has restored to His Messenger from the people of the towns, it is for Allah and for the Messenger, and for the near of kin and the orphans and the needy and the wayfarer, so that it may not be a thing taken by turns among the rich of you, and whatever the Messenger gives you, accept it, and from whatever he forbids you, keep back, and be careful of (your duty to) Allah; surely Allah is severe in retributing (evil).*

Muadh ibn Jabal, a companion of the Prophet, had this to say about the immunity of the executive in power: "Our leader is one of us . . . if he commits theft, we shall amputate his hand; if he commits adultery, we shall flog him. . . . He will not hide himself from us, nor will he be self-conceited. . . . He is a person as good as we are." According to the Prophet a ruler is just a common man, voted into power. He is not exempt from punishment or impeachment when his actions are examined in a legitimate judicial process.

So if these ideas come from Islamic teachings, it may seem inexplicable that they haven't been in place in Muslim governments. Democracy, as we know it, is the expression of the will of the people. In earlier generations, when populations were small, there was no need for casting of the ballot. The views of the people were aired in public meetings, generally in a mosque, a place that the public regularly visited. Muslim jurists did not know the formal word "democracy"; rather, democracy was a consensus emerging from the free expression of the will of the small community as well as the implementation of quranic ideals that we know to be intrinsic to democratic society: tolerance, pluralism, justice, law, equality, and a fair economic system.

I think it's pretty clear that you have to go a long way, and deliberately exclude quranic text and Muslim history, to try to make an intellectual case that Islam and democracy are mutually exclusive and can't coexist. Indeed, it is dictatorship that is abhorrent and prohibited under Islam. But an even harder stretch of theology is required to justify or rationalize terrorism, suicide bombings, or any violence directed at the innocent within the Islamic tradition or within the divine words of the Holy Book.

.

Some people assert that democracy will not work in an Islamic country because Muslims believe in the sovereignty of God and thus cannot accept man's law. God is Master of the Universe, of the known and unknown. Humans share two relationships: one with God and one with one another. They are custodians of God's trust, the earth, which has been placed in their care, as they are created by God. God has sent his principles to humans through thousands of Prophets, including Moses, Abraham, Jesus, and Mohammad (who is the last messenger), to instruct us how we should conduct our lives and the principles by which we should conduct our societies. The immutable principles of justice, truth, and equality must not be transgressed if we are to gain entrance to everlasting life in Paradise.

Thus humans must seek and apply knowledge, must use reason, must consult and build a consensus for a just society on earth on which they will be judged in the hereafter. They must not sin by taking innocent life, for God alone has the right to give and take life. Anyone who interferes in God's work by taking a life commits the most heinous crime in Islam.

The terrorists who attacked me with two bomb blasts on October 19, 2007, when I returned to Pakistan to a historic reception, committed the most heinous crime of murder by taking the lives of 179 innocent people. So too does anyone who attacks innocent people, whether in the World Trade Center, the tubes in London, or the resorts of Bali, Indonesia.

I am told that the terrorists who made the bombs and conspired to kill me took a *fatwa,* or religious edict, to sanctify the terrorist attacks. However, on the Day of Judgment, such an edict will be of no help. God has ordained that each individual will have to account individually for his actions without intercession from any other individual.

Under the Constitution of Pakistan, authored by my father and passed unanimously by Pakistan's Parliament in 1973, the democratic right to Muslim governance is recognized. The Constitution of 1973 states, in its preamble:

Whereas sovereignty over the entire Universe belongs to Almighty Allah alone, and the authority to be exercised by the people of Pakistan within the limits prescribed by Him is a sacred trust;

And whereas it is the will of the people of Pakistan to establish an order:

Wherein the State shall exercise its powers and authority through the chosen representatives of the people;

Wherein the principles of democracy, freedom, equality, tolerance and social justice, as enunciated by Islam, shall be fully observed;

Wherein the Muslims shall be enabled to order their lives in the individual and collective spheres in accordance with the teachings and requirements of Islam as set out in the Holy Quran and Sunnah;

Wherein adequate provision shall be made for the minorities freely to profess and practise their religions and develop their cultures;

Wherein the territories now included in or in accession with Pakistan and such other territories as may hereafter be included in or accede to Pakistan shall form a Federation wherein the units will be autonomous with such boundaries and limitations on their powers and authority as may be prescribed;

Therein shall be guaranteed fundamental rights, including equality of status, of opportunity and before law, social, economic and political justice, and freedom of thought, expression, belief, faith, worship and association, subject to law and public morality;

Wherein adequate provision shall be made to safeguard the legitimate interests of minorities and backward and depressed classes;

Wherein the independence of the judiciary shall be fully secured;

Wherein the integrity of the territories of the Federation, its independence and all its rights, including its sovereign rights on land, sea and air, shall be safeguarded;

So that the people of Pakistan may prosper and attain their rightful and honoured place amongst the nations of the World and make their full contribution towards international peace and progress and happiness of humanity:

Now, therefore, we, the people of Pakistan,

Cognisant of our responsibility before Almighty Allah and men;

Cognisant of the sacrifices made by the people in the cause of Pakistan;

Faithful to the declaration made by the Founder of Pakistan, Quaid-i-Azam Mohammad Ali Jinnah, that Pakistan would be a democratic State based on Islamic principles of social justice;

Dedicated to the preservation of democracy achieved by the unremitting struggle of the people against oppression and tyranny;

Inspired by the resolve to protect our national and political unity and solidarity by creating an egalitarian society through a new order;

Do hereby, through our representatives in the National Assembly, adopt, enact and give to ourselves, this Constitution.

Thus we can see that there is a perfect constitutional template for democratic governance in the Muslim world. But the current poor relations between much of the West and much of the Islamic world may suggest the need for new terminology if we are to realize the vision. The word "secular," used to denote separation of state and religion in the Western world, often means "atheism," or rejection of God, when translated into other languages, including into Urdu in Pakistan.

Instead of terms such as "secularism," the director of the Study of Muslim Civilizations at the Aga Khan University in London, Dr. Abdou Filali-Ansary, believes that we should refer directly to the individual building blocks of democracy—free elections, an independent judiciary, respect for women's and minority rights, the rule of law, and fundamental freedoms—to describe the true meaning of a democratic society. We shouldn't be talking secularism, which to Muslims is a clouded, misleading, and sometimes contentious term. Instead of using terms that fall into the rhetorical trap set by extremists to discredit the elements of modern democratic society, we should rather stress elements such as freedom to travel, freedom to work, opportunity for education for both sexes, the independence of the judiciary, and a robust civil society. These issues, more than the term "secularism," connote the compatibility of Islam and democratic values.

Who can doubt that Islam—as a religion and as a value structure—has been distorted and manipulated for political reasons by militants and extremists and dictators. The establishment of the Afghan mujahideen by Zia in the 1980s is an example. (After all, the jihad in Afghanistan aimed to rid the country of Soviet occupation, not reject modernity, technology, and pluralism, and to establish "strategic depth" in Pakistan. That was a political agenda of Zia.) Islam is now being used for purely political purposes by a group of people who are angry with the West. Religion is being exploited and manipulated for a political agenda, not a spiritual agenda.

The militants seethe with anger, but their anger is always tied to their political agenda. First, they were angry that the West had abandoned three million Afghan refugees and stopped all assistance to them after the Soviets left Afghanistan. Second, they are angry that their offer to the government of Pakistan to send one hundred battle-hardened mujahideen to help in the Kashmir uprising of 1989 was rejected. Third, they wanted King Fahd of Saudi Arabia to turn to their "battle-hardened mujahideen" to protect Saudi Arabia after Iraqi president Saddam Hussein invaded Kuwait on August 2, 1990. He refused. Fourth, they went off to fight in Bosnia when the region was engulfed in war (from 1993 to 1996 I lobbied President Bill Clinton, Prime Minister John Major, and other European leaders to intervene to bring the conflict to an end). Fifth, they tried to exploit the Chechen nationalist movement. Sixth, with the fall of my government, they turned their attention to Kashmir and tried to take over the nationalist Kashmiri movement from 1997 onward.

Muslim extremists systematically targeted historical nationalist movements to gain credibility and launch themselves into the Muslim heartland with a view to piggybacking off nationalist movements to advance their agenda. However, most Muslims were suspicious and not welcoming of their extreme interpretation of Islam. Thus it was only in Afghanistan, already softened by years of resistance by Afghan mujahideen, that Muslim extremists were able to establish the Taliban dictatorship.

Driven out of Afghanistan after the September 2001 attacks on the United States, they returned to Pakistan, where the journey had begun with General Zia-ul-Haq in 1980.

After the United States invaded Iraq, these same extremists turned their attention to that country. Abu Musab al-Zarqawi went off to fight in Iraq. Presumably others did, too. Again they used religious propaganda to kill, maim, and effectively divide one of the richest Muslim countries, Iraq, into a land of carnage and bloodshed. Sunnis and Shias, who had lived peacefully side by side for centuries, began to kill each other, and Iraq began to fall apart. It is quite easy (and typical) for Muslim extremists to blame the Americans for the sectarian civil war that rages in Iraq today, when actually it is a long-standing tension between Muslim communities that has been exacerbated and militarized to create the chaos under which extremists thrive.

Iraq is not the only goal of the extremists. Pakistan too is in great danger. Pro-Taliban forces have taken over the tribal areas of Pakistan. They occupy the Swat Valley. They have been ceded Waziristan by the Musharraf regime. They are moving into the settled areas of Pakistan. Their apparent next goal is the cities of my country, including our capital, Islamabad. They thrive on dictatorship; they thrive on terror; they provoke chaos to exploit chaos.

I returned to Pakistan on October 18, 2007, with the goal of moving my country from dictatorship to democracy. I hoped that this transition could take place during the scheduled elections of 2008. I feared that otherwise the extremists would march toward Islamabad. Islamabad is near the town of Kahuta, where Pakistan's nuclear program is being carried out.

It is my fear that unless extremism is eliminated, the people of Pakistan could find themselves in a contrived conflict deliberately triggered by the militants (or other "Islamists") who now threaten to take over Pakistan's nuclear assets. Having a large Muslim nation fall into chaos would be dangerous; having the only nuclear-armed Muslim nation fall into chaos would be catastrophic. My people could end up being bombed, their homes destroyed, and their children orphaned simply because a dictator has focused all his attention and all of the nation's resources on containing democrats instead of containing extremists, and then has used the crisis that he has created to justify those same policies that caused the crisis. It may sound convoluted, but there is certainly method to the madness.

This is such a tragedy, especially because Islam is clearly not only tolerant of other religions and cultures but internally tolerant of dissent. Allah tells us over and over again, through the Quran, that he created people of different views and perspectives to see the world in different ways and that diversity is good. It is natural and part of God's plan. The Quran's message is open to and tolerant of women's full participation in society, it encourages knowledge and scientific experimentation, and it prohibits violence against innocents and suicide, despite terrorists' claims to the contrary.

Not only is Islam compatible with democracy, but the message of the Quran empowers the people with rights (democracy), demanding consultation between rulers and ruled (parliament), and requiring that leaders serve the interests of the people or be replaced by them (accountability).

Islam was sent as a message of liberation. The challenge for modern-day Muslims is to rescue this message from the fanatics, the bigots, and the forces of dictatorship. It is to give Muslims back the freedom God ordained for humankind to live in peace, in justice, in equality, in a system that is answerable to the people on this earth accepting that it is God who will judge us on the Day of Judgment.

It is by accepting that temporal and spiritual accountability are two separate issues that we can provide peace, tranquillity, and opportunity. There are two judgments: the judgment of God's creatures in this world through a democratic system and the judgment by God when we leave this world. The extremists and militants who seek to hijack Islam aim to make their own judgments. In their failure lies the future of all Muslims and the reconciliation of Islam and the West.

3

Islam and Democracy: History and Practice

Conventional wisdom would have us believe that democracy has failed to develop in the Muslim world because of Islam itself. According to this theory, somehow Islam and democracy are mutually exclusive because Islam is rooted in an authoritarianism that promotes dictatorship. I reject this thinking as convenient and simplistic, grounded in neither theology nor experience. As a Muslim who has lived under both democracy and dictatorship, I know that the reasons are far more complex.

The so-called incompatibility of Islam and democratic governance is used to divert attention from the sad history of Western political intervention in the Muslim world, which has been a major impediment to the growth of democracy in Islamic nations.

The actions of the West in the second half of the nineteenth century and most of the twentieth century often deliberately blocked any reasonable chance for democratic development in Muslim-majority countries. It is so discouraging to me that the actions of the West in the pursuit of its various short-term strategic goals have been counterproductive, often backfiring. Western policies have often preserved authoritarianism and contained the growth of nascent democratic movements in the developing world, specifically in the Islamic world. Western nations' efforts to disrupt democratic tides—initially for economic reasons and then for political ones—have fueled and exacerbated tensions between the West and Islam.

Despite often grand rhetoric to the contrary, there has been little real Western support for indigenous democratic movements. Indeed, too often there has been outright support for dictatorships. Both during the Cold War and now in the current battle with international terrorism, the shadow between Western rhetoric and Western actions has sowed the seeds of Muslim public disillusionment and cynicism. The double standards have fueled extremism and fanaticism. It accounts, at least in part, for the precipitous drop in respect for the West in the Muslim world. This trend is true even in pro-Western Muslim countries such as Turkey. When I was growing up, I thought of Western nations as inspirations for freedom and development. I still do, but I'm afraid I'm in a shrinking minority of Muslims.

There is an abundance of other examples that manifest the inconsistency of Western support for democracy in the Muslim world: specifically, Western actions that undermined democratic institutions, democratic movements, and democratically elected governments in countries that the West considered critical to other policy objectives. The countries range from large to small, from very important to relatively insignificant. What is remarkable is the clear pattern of Western action: perceived pragmatic self-interest trumping the values of democracy, almost without exception. In a nation that is not relatively strategically important, such as Burma, the West will enforce its democratic creed quite enthusiastically, organizing trade embargoes and other forms of political isolation. But in places that are viewed as strategically important for economic or geopolitical factors, the West's commitment to democracy can often be more platitude than policy.

I raise this as not just a strategic inconsistency but a true moral dilemma for the West, especially the United States. On one level the West speaks of democracy almost in the context of the values of religion, using rhetoric about liberty being a "God-given" right. And Western nations often take that standard abroad, preaching democratic values like missionaries preaching religion. The problem arises, of course, in its selective application to bilateral foreign policy relationships. I have always believed, and have publicly argued, that the selective application of morality is inherently immoral.

If dictatorship is bad, then dictators are bad—not just dictators who are impotent and irrelevant but also those who are powerful allies in fighting common enemies. The West makes human rights the centerpiece of its foreign policy selectively. The West also stands foursquare with struggling democracies selectively. In his second inaugural address, President George W. Bush said:

> *We will encourage reform in other governments by making clear that success in our relations will require the decent treatment of their own people. America's belief in human dignity will guide our policies, yet rights must be more than the grudging concessions of dictators; they are secured by free dissent and the participation of the governed. In the long run, there is no justice without freedom, and there can be no human rights without human liberty.*

President Bush's words notwithstanding, Washington supported Pakistan's military dictator, General Musharraf, whom it considered a key ally in the war against terrorism, even as it simultaneously supported democracy in neighboring Afghanistan and in Iraq in the Middle East.

I am not the only one, of course, who has pointed to strategic and moral inconsistencies in the application of Western political values abroad. Recently Noah Feldman wrote in *The New York Times Magazine* that "a republic that supports democratization selectively is another matter. President Bush's recent speech to the United Nations, in which he assailed seven repressive regimes, was worthy of applause—but it also opened the door to the fair criticism that he was silent about the dozens of places where the United States colludes with dictators of varying degrees of nastiness." Feldman specifically cites my homeland of Pakistan as one example but goes on to criticize

American support for Hosni Mubarak of Egypt as Mubarak cracks down on the press and other political parties. Feldman adds that "Saudi Arabia—one [of the United States'] most powerful and durable allies—hasn't moved beyond the largely symbolic local council elections that it held two years ago." The United States, berating Burma and Iran for their undemocratic brutality, has had little to say about U.S. allies. Again, the selective application of morality is criticized as immoral in many nations whose people are also striving for democracy.

There is a clear relationship between dictatorship and religious fanaticism that cannot be ignored. Carl Gershman, the president of the National Endowment for Democracy, has referred to it as a relationship between autocrats and the Islamists. To the extent that international support for tyrannies within Islamic states has resulted in the hostility of the people of these countries to the West—and cynicism about the West's true commitment to democracy and human rights—some might say that the West has unintentionally created its own Frankenstein monster.

I cannot dispute that there have been few sustained democracies in the Islamic world. But the responsibility does not lie in the text of the Muslim Holy Book. It is a responsibility shared by two significant elements that have come together in the context of environmental conditions inhospitable to the establishment, nurturing, and maintenance of democratic institutions in Muslim-majority societies.

The first element—the battle within Islam—is the purportedly theological fight among factions of Islam that also often seeks raw political and economic power at the expense of the people. The second element—the responsibility of the West—includes a long colonial period that drained developing countries of both natural and human resources. During this time the West showed a cold indifference toward supporting democracy among Muslim states and leaders for reasons that were either economic (oil) or political (anticommunism).

We cannot minimize the fault line that has existed within Muslim nations, a fault line of internal factionalism, disrespect for minority rights, and interventionist and often dysfunctional military institutions. These elements have often been accompanied by the presence of authoritarian political leadership. There is obviously a shared responsibility for democracy's weakness in Muslim-majority states, but there can be no disputing the fact that democratic governance in Muslim countries lags far behind that in most other parts of the world.

.

A useful context for the history of democracy within Muslim countries is provided by a brief review of current categorizations of political rights and civil liberties around the world. It will then be possible to objectively compare the Muslim and non-Muslim worlds on standards and criteria of democratic development. Central to this analysis is something that I have always believed and strongly endorse: that freedom and liberty are universal values that can be applied across cultures, societies, religions, ethnic groups, and individual national experiences. Democracy is not an inherently Western political value; it is a universal value. Liberty means as much to someone from Indonesia as it does to someone from Louisiana.

Freedom House (which was founded at the beginning of World War II by First Lady Eleanor Roosevelt and Wendell Willkie, the Republican candidate whom her husband had just defeated for the presidency) is an international nongovernmental organization (NGO) dedicated to promoting democracy, human rights, and freedom around the world. Each year it engages scholars from around the world to categorize governments on a scale of political rights ranging from "totally free" to "not free." This useful analytical tool is based on analyses of electoral processes, political pluralism and participation, and how the government functions. Countries are scored on a numerical scale that ranges from one to seven, with the highest number representing the lowest level of freedom. This number is then used to determine one of three ratings: free, partly free, or not free. In some cases, additional variables are used to supplement the data. For example, for traditional monarchies international scholars are additionally asked if the system provides for genuine, meaningful consultation with the people, encourages public discussion of policy choices, and permits petitioning the ruler.

The analysis is especially useful in evaluating political systems in predominantly Muslim monarchies, because it integrates the elements of legitimate secular government with the citizen consultation enshrined in the Quran. The disparities in Freedom House ratings between the Muslim and non-Muslim worlds are dramatic and statistically significant, but not particularly surprising. It is important to remember, of course, that Muslim nations are very different from Western nations in national experience. Specifically, Islamic law generally has a role in government, whether in secular Islamic states such as Kazakhstan or religiously ideological countries such as the Islamic Republic of Iran.

Of the forty-five predominantly Muslim states, only Indonesia, Mali, and Senegal are considered free. Eighteen Muslim nations are considered partly free: Afghanistan, Albania, Bahrain, Bangladesh, Comoros, Djibouti, Gambia, Jordan, Kuwait, Kyrgyzstan, Lebanon, Malaysia, Mauritania, Morocco, Niger, Sierra Leone, Turkey, and Yemen.

Twenty-four predominantly Muslim nations are labeled not free: Azerbaijan, Brunei, Egypt, Palestine, Guinea, Iran, Iraq, Jordan, Kazakhstan, Libya, Maldives, Oman, Pakistan, Qatar, Saudi Arabia, Somalia, Sudan, Syria, Tajikistan, Tunisia, Turkmenistan, United Arab Emirates, Uzbekistan, and Western Sahara.

The mean score for political rights (on a scale of 1 to 7, with 1 being the highest level of rights) in the Muslim world is 5.24, compared to 2.82 for the non-Muslim world. The mean score for civil liberties in Muslim countries is 4.78, compared to 2.71 for non-Muslim countries. These are significant differences. I believe that these differences are not the result of theology but rather a product of both Western manipulation and internal Muslim politicization of Islam.

·

One frequently overlooked detail in the analysis of Freedom House scores is the difference between Arab and non-Arab Muslim-majority countries. In "An 'Arab' More Than 'Muslim' Electoral Gap," Alfred Stephan and Graeme Robertson use two different indices of levels of political rights to compare these two types of Muslim-majority countries. The study contrasts the scores of countries in the Freedom House study and also in the Polity IV Indexes relative to GDP from 1972 to 2000, when the competitiveness of an election was questioned. (The Polity IV Project codes and compiles information on the regulation and competitiveness of political participation.)

The authors differentiate between "underachievers" and "overachievers" in electoral competitiveness, defined by such criteria as whether the government was selected by reasonably fair elections and whether the democratically elected government actually wields political power.

Stephan and Robertson found that a non-Arab Muslim-majority country was astoundingly "almost 20 times more likely to be 'electorally competitive' than an Arab Muslim-majority country." Of the forty-seven Muslim-majority countries that they studied, the Arab Muslim countries formed "the largest single readily identifiable group among all those states that 'underachieve,' " but the world's thirty-one Muslim-majority non-Arab countries form the largest bloc that "greatly overachieves" in electoral competitiveness. In studying the thirty-eight countries in the world that suffer from extreme poverty, they found "no comparative Muslim gap whatsoever when it comes to political rights." Their findings suggest that the success of democracy within certain states has less to do with whether a country has a Muslim majority than was previously thought by Western analysts. The result shatters the hypothesis that religion is a key variable related to democracy and that Islam and democracy are inconsistent. It relegates the Islam-democracy incompatibility theory to the level of mythology.

.

Democracies do not spring up fully developed overnight, nor is there necessarily a bright line between democratic governance and autocracy. More typical, democracy can be seen on a continuum. Civil society and democratic institutions such as political parties and NGOs tend to develop slowly over time, one critical step at a time.

True democracy is defined not only by elections but by the democratic governance that should follow. The most critical elements of democratic governance go beyond just free and fair elections to the protection of political rights for those in political opposition, the open function of a civil society and free press, and an independent judiciary. Far too often in the developing world—including the Islamic developing world—elections are viewed as zero-sum games. The electoral process is democratic, but that's where democracy ends. What follows is tantamount to one-party authoritarian rule. This is the opposite of true democratic governance, which is predicated on shared constitutional power and responsibility. And because democratic governance rests on a continuum of experience, the length of that experience is directly related to the sustainability of democratic governance itself. In other words, the longer democratic governance is maintained, the stronger the democratic system becomes.

A democracy that is more than two hundred years old is not in serious danger of interruption or of suspension of constitutional norms. It has a two-century-old firewall of democratic history and practice to protect itself from extraconstitutional abuse of power. A nation without such a long history of democracy and democratic institutions—political parties; a popularly elected, legitimate, sovereign parliament; NGOs; free media; and an independent judiciary—is vulnerable to the suspension of the democratic order. We must think of a new democracy like a seedling that must be nourished, watered, fed, and given time to develop into a mighty tree. Thus, when democratic experiments are prematurely interrupted or disrupted, the effects can be, if not permanent, certainly long-lasting. Internal or external interruptions of democracy (both elections and governance) can have effects that ripple and linger over generations.

We must be realistic and pragmatic about democracy. John F. Kennedy once referred to himself as an "idealist without illusions." To me this is a useful description as I think in particular of my country moving from the brutality of dictatorship to the civility of democracy. When confronted with tyranny, one is tempted to go to the barricades directly, when pragmatism would dictate exhausting other potential (and peaceful) remedies. As I have grown in maturity and experience, I remain as strongly committed to the cause but more patient in finding means to achieve goals peacefully.

The colonial experience of many Muslim countries had contributed to their difficulties in sustaining democracy. In the absence of adequate support and without the time and commitment needed to build a democratic infrastructure, they failed to strengthen their electoral and governing processes. Many of the countries discussed in this chapter were exposed to democratic values, democratic ideals, and the gradual development of political and social institutions while under colonial rule or shortly thereafter. However, their nascent democratic seeds were often smothered by the strategic interests of Western powers (often working with elements within their own societies) before they flowered into viable democratic systems.

Iran

Iran may be the seminal case of one of my principal theses, for its consequences continue to have a major impact across the entire Islamic world and potentially threaten the stability of regional and global international relations. Iran is a clear example of the West trumping basic democratic and human values to address its short-term political and economic goals. It is a pattern that has repeated itself with tragic regularity throughout the better parts of two centuries.

Iran was ruled by a succession of shah dynasties from 1848 to the Khomeini revolution of 1979. The most ironic element of the Iran story is that parliamentary democracy was establishing a strong foothold in Tehran in the early 1950s. The process was disrupted by a British-American collaboration that resulted in an August 1953 CIA-sponsored coup against a legitimate, democratically elected Iranian government.

This coup not only destroyed democratic governance in Iran but made generations of Muslims suspicious and cynical about Western motivations. It heightened sensitivity to the gap between Western rhetoric and Western actions. And since the CIA-sponsored action was covert, the American people were not even aware of the actions of their government.

For a long time Persia (later Iran) was a poster child of resource exploitation by colonial powers. Russia manipulated and occupied the country, depleting its resources. As far back as 1891, the British seized control of the entire Iranian tobacco industry. This provoked the first public insurrection that ultimately forced the then shah to recognize the power of protest and to become sensitive to what a century later would be called "the street." The Iranians felt that the British had paid a pittance for the full rights to Iranian tobacco, breeding resentment.

Ten years later, an Iranian leader, desperate for cash, brokered away the future of his country. He sold the rights to all the natural gas and petroleum in Persia to the Englishman William Knox D'Arcy. Although the sale took place before oil had been discovered in Iran, it provided the legal basis for Britain's claim to Iran's oil reserves. Shortly thereafter—and partly because of the deal—a spontaneous groundswell of protest and demand for political and economic reform broke out. The shah reluctantly agreed to the establishment of a legislature, modeled on the European parliamentary system. It was the vanguard of a heroic fight by the people of Iran for democracy and control over their own lives and their nation's wealth.

Despite the nascent elements of representative government, the newly created Parliament could not block the collusion of the shah with imperial powers the following year, resulting in a functional partition of Iran. Russia controlled the north and Britain the south, leaving the Iranian people with only a small strip in the middle. The monarch, Mohammad Ali Shah, undermined the independence of the new Parliament with the backing of both his British and Russian partners. Nevertheless, the Parliament proceeded on an independent path, reasserting its legitimacy and popular support. In 1911 the independent Iranian Parliament hired an American named William Morgan Shuster as Iran's treasurer general. Shuster's mandate was to dismantle the financial system that enabled imperial powers to control the wealth of the nation with impunity. One of those imperial powers, Russia, actually sent troops into Tehran to shut down Parliament.

The constitutional developments taking place between 1905 and 1911 were in effect a revolution marking the end of the feudal system in Iran. That feudal system was contradictory to the Islamic principles of consultation and equality. The Iranian people created a written constitution in a bid to establish a new social and political order. This historic moment marking the beginning of democratic institutions was significant not just for Iran but potentially for the entire Muslim world.

The emergence of democratic institutions threatened the sovereignty and authority of the shah's rule. It made him accountable to the people. It also challenged the special relationship between the foreign powers of Britain and Russia and the shah, who gave them direct access to Iran's natural resources. The destabilization of a legitimately elected popular government enjoying the popular support of its people through collusion between an authoritarian Islamic ruler and Western powers would be repeated throughout the Muslim world. It was the beginning of a recurring nightmare for democracy in Iran, a clear pattern of foreign political intervention to support Western economic and political goals. It would serve as an unfortunate precedent for similar collaborations between regressive Islamic authoritarian regimes and foreign powers throughout the twentieth century and the beginning of the twenty-first century.

In 1911 the Anglo-Persian Oil Company (later the Anglo-Iranian Oil Company [AIOC]) began pumping oil out of Iran. In 1917 the Bolshevik Revolution resulted in Russia's renouncing its claims to Iran, allowing Britain to move in and take over the Russian interests and areas of control. In 1919 the British signed the Anglo-Persian Agreement with Ahmad Shah, which essentially made Iran a British protectorate, giving it control of Iran's military, transportation system, communications, and treasury.

Only two years later, the British helped install a new strongman with whom they could work closely to control Iran. His name was Reza Khan. Reza Khan signed the 1933 agreement between the Anglo-Persian Oil Company (controlled by Britain) and the government of Iran, providing a small annual stipend to Iran and huge profits for the Anglo-Persian Oil Company. The unholy alliance between Britain and Reza Khan did not last long, however, since Khan gave enthusiastic assistance to the Nazis during World War II. His collusion with the Axis powers caused the British and Soviets to occupy Tehran and force him to abdicate. On Reza Khan's abdication, his son Mohammad Reza Shah assumed the Peacock throne.

In 1942 the first Iranian political party, Tudeh, was established. From the outset, Tudeh challenged the shah for control of the direction of the country. The Tudeh nationalists were much more independent of British economic control than the pliable shah was. In 1949, after an assassination attempt against the shah (which may have been contrived to tighten control over the political system), the shah banned the Tudeh party and rigged parliamentary elections. This led to protests led by Mohammad Mossadegh, whose supporters formed the National Front, a broad political coalition dedicated to increasing democracy and limiting foreign power in Iran. Mossadegh would become one of the central figures of the emerging Iranian nationalist and democratic movements and later the centerpiece and target of

Operation Ajax, which set the cause of democracy in Iran back generations.

In 1950 the British, worried they were losing control, forced the shah to name a pro-British general, Ali Razmara, as prime minister. A year later, Razmara was assassinated and Parliament almost immediately voted to nationalize Iran's oil. Mossadegh, the head of the National Front, gained enormous national support and was elected prime minister, threatening British interests in Iran. In August 1951 Britain imposed a full-scale trade embargo on Iran. Additionally, the British explored covert ways to get rid of Mossadegh, who continued to assert his power and authority, including challenging the shah for control of the Iranian military. When the shah tried to rein him in, Mossadegh became obdurate. He pressed the shah with demands for political democratization and economic reform. When the shah rebuffed him, Mossadegh resigned.

The new prime minister, a tool of both the shah and Britain, began to renegotiate British control over Iran's nationalized oil resources. Huge street protests broke out with religious groups participating. Mossadegh was reinstated, and, owing to his enormous popular support, he gained control over the military. Soon thereafter, rumors of a planned coup began to circulate. Mossadegh reacted by breaking diplomatic ties with Britain. He forced the removal of all British intelligence personnel from Iran. Mossadegh's nationalism and his willingness to stand up to Britain made him a national hero amongst the people of Iran. But by alienating Britain, Mossadegh paid a tremendous political price with huge ramifications for Iranian history.

The intrigue in Tehran coincided with the 1952 presidential election in America. The U.S. election was taking place against the background of the Cold War with the Soviets and the hot war with the Chinese in Korea. Soon after Dwight D. Eisenhower was elected president in November, the British government sent a top intelligence officer to Washington.

A CIA agent, Monty Woodhouse, informed U.S. intelligence

about the British assets remaining in Iran, the political situation in the country, the economic vulnerability of Iranian oil, and the unreliability of Prime Minister Mossadegh in the effort to contain international communism (Iran, of course, borders on the former Soviet Union). Woodhouse met with the Central Intelligence Agency in Washington, D.C. The CIA was responsive to his suggestion that a political and economic disaster for the West was unfolding in Iran. Woodhouse and the CIA began plotting alternative strategies for U.S. covert action against the Mossadegh government.

Convinced by the CIA that Mossadegh was a pro-Soviet and anti-American leader, President Eisenhower gave the green light to Operation Ajax on June 25, 1953. Through a combination of paid political ads, paid street agitators, fomented street protests, and a rebellion triggered by pro-shah military leaders, Kermit Roosevelt, director of the CIA's Middle East Bureau (and the grandson of U.S. president Theodore Roosevelt) devised Operation Ajax. Operation Ajax aimed to and actually did incite chaos all across Iran. The anarchy and turmoil created a political environment that allowed the shah to sack the Mossadegh government. The shah, working with the CIA, had loyal military units physically seize the government and install Fazlollah Zahedi as the new prime minister. Democracy in Iran had been effectively snuffed out. And it has yet to return.

Thus, starting in 1919, Britain treated Iran as a colonial state. As the sun set over the British Empire, the British relied increasingly on Iranian oil to maintain their stature as a world power. Since the British government owned a controlling interest in the AIOC, they were able to use the Iranian oil to fuel the Royal Navy at almost no cost.

Britain turned to Washington to protect its interests and act as its political "muscle." British officials, who had failed to convince the Truman administration of the need for a tough stance against Mossadegh, succeeded in convincing the new, ideologically rigid anti-Communist Eisenhower administration of the "Communist threat" that Mossadegh posed. At the height of the Cold War, the British appeal to the Eisenhower administration had significant resonance. The United States could not afford another Communist "domino" falling to the Soviets, especially one dripping in oil reserves. Painting the democratically elected nationalist Mossadegh government as Communist was an effective psychological ploy in international relations of the time.

This phenomenon of viewing nationalists as Communists often became a self-fulfilling prophecy. One can only imagine how different the world might be if the French had embraced Ho Chi Minh the nationalist leader before their extended war against him turned him into a Communist despot. Further, many scholars believe that if the United States had used its influence in Cuba to attempt to liberalize the Fulgencio Batista regime and to aid the new Fidel Castro regime after the revolution in 1957, Castro might have taken a different ideological turn. And even after the Cuban Revolution succeeded, if the United States had offered economic incentives and economic assistance to the young Castro regime, many believe Fidel Castro's "conversion" to communism might have been avoided.

After the American-led coup, the shah moved to consolidate and secure his power. He strengthened the military, buying more than $10 billion worth of defense equipment from the United States from 1952 to 1956 alone. The shah purged his military of pro-Mossadegh elements and arrested scores of Tudeh members, executing twenty-seven of them. Democratic leaders were viewed as seditious.

In 1957, with the active help of the United States, the shah formed SAVAK, a brutal internal security agency that fought internal dissent and eliminated opposition. SAVAK became a synonym for dictatorial terror, and it became well known to the Iranian people that the guns in SAVAK's hands and the methods it employed were products of the shah's American benefactors. The shah, like so many dictators after him, played the West, and in particular the United States, like a fiddle. Throughout his quarter-century rule, he effectively exploited America's fear of communism to achieve his goals of building up his military and destroying political opposition, consolidating his power and choking off his enemies. If the United States balked, the shah threatened to go to the Soviets for help.

This pattern of dancing with anti-Communist dictators was a hallmark of the West's international policies for a half century, stifling democratic development in countries in Asia, Africa, Europe, and Latin America. The courtship of brutal authoritarians by successive U.S. administrations helped to create, by the late twentieth century, a successor generation of democratic political leaders in the newly independent world that was suspicious of Western political motivation and sensitive to the shadow—which they saw as hypocrisy—between the West's rhetoric and action.

From 1953 onward, the shah became increasingly repressive. In 1955 he deposed Prime Minister Zahedi and took all power into his own hands. When the debate around the 1960 elections brought too many charges of corruption, he summarily canceled the elections. Following protests, he relented and installed a new pro–land reform prime minister, but he never followed through on the proposed reforms.

The opposition gathered momentum in 1963 after the government killed a prominent religious leader, resulting in massive protests led by the young Ayatollah Ruhollah Khomeini. The shah ordered his army to shoot at the protestors. Thousands of people died. Khomeini was arrested and exiled. The shah convinced the United States that this violence had been caused not by legitimate protestors but by revolutionaries and Communist agitators and sympathizers. Once again, acting out of the "self-interest" Cold War model, the United States elected to have its Soviet-containment strategy trump its commitment to basic democratic values. The geopolitical strategy had little resonance in Iran. The Iranian people perceived the United States to have turned a blind eye to repression and sided with a dictator in expanding authoritarian control. The 1963 elections that followed were a sham, coalescing the shah's power and eliminating all opposition.

Years of repression finally crumbled in 1979. While most Americans did not know of their country's hand in the events of 1953 and onward, this history was well known within Iran and passed on to a new generation. When President Jimmy Carter allowed the shah to enter the United States for cancer treatment after the 1979 coup, many in Iran had flashbacks to the totally unrelated U.S. reinstatement of the shah in 1953 after he fled to Rome. In their view a quarter century of America's dancing with the shah—supporting his regime, despite its tyranny, as long as economic interests (oil) and political interests (containing communism)—was maintained. This policy had tragic consequences not only for the people of Iran but also for the long-term interests of the West.

The Khomeini revolution can be traced to the shah's resistance to reform, the West's not pressing for reform, and to the West's intervention against a democratically elected Parliament and prime minister in 1953. A covert CIA coup against democracy aborted the development of democratic infrastructure and civil society—political parties, NGOs, a legitimate and functioning parliament, an independent judiciary, an uncensored media—in Iran. These basic building blocks of democracy and pluralism undermine religious extremism and fanaticism. It is important to note that Iran was one of the least extremist of the Muslim nations. Despite the shah's autocratic rule, social tolerance was widespread, and exposure to information and technology was possibly greater in Iran than anywhere else in the Islamic world. Had democratic institutions been allowed to develop, they might well have insulated Iran from the Khomeini revolution.

While the people of Iran were less likely than others in the Islamic world to opt for an extremist religious future, years of repression and the suppression of support for democracy had a tremendous impact on Iranians' political attitudes. When confronted with tyranny by a dictator supported by the West, the "street" turned extremist, resulting in the Khomeini revolution. Liberalization would have made this totally avoidable. In 1979 a religious regime replaced an autocratic secular regime that had replaced a democratically elected government in 1953. The lesson of Iran can be applied today in Pakistan and other

Muslim nations. Religious extremism is directly related to the suppression of individual liberty and to dictatorial rule, supported by Western governments to further their own short-term political agenda. Whether it's in the name of the "global war on terror," the containment of communism, or protecting access to scarce natural resources, selective application of democratic principles creates long-term dangers and consequences.

If one believes, as I do, that liberty is a universal value to which all people aspire, acquiescence to autocracy is not acceptable. In most cases, transition from colonial rule to independent rule had little immediate effect on the individual liberty of citizens within a society or on their day-to-day lives. National independence had more economic consequences than political—after independence, resources were less likely to be exploited from abroad. Unfortunately, even after independence, the national economic benefits were not distributed equitably within society. Often one tyranny replaced another, sometimes even more brutal and unforgiving than what had preceded it. This ugly pattern is partly a consequence of the colonial experience, in which Western democratic nations refused to reconcile the democratic values that governed them at home with the undemocratic conduct that characterized their behavior abroad.

But that is only part of the explanation. Undeniably, a significant cause of the lack of democratic development in Muslim-majority nations rests within these countries themselves. Factionalism and the battle between moderation and extremism often derailed the natural progression to democracy. The exploitation and brutalization of Muslim masses and Muslim elites by Muslim dictators are stains on the Muslim community that cannot easily be explained away.

Algeria

The North African Muslim nations are examples of countries where independence did not bring the people political liberty or economic and social reforms. For example, Algeria was under French colonial domination for more than a hundred years.

The anticolonial movement in Algeria grew between the world wars and accelerated with the defeat of the Nazis in 1945. As the war concluded, an anticolonial parade in the town of Setif turned violent. Algerians attacked the French and European communities. The death toll exceeded one hundred. The French forces responded at first with appropriate and restrained action but soon turned to reprisal killings; many thousands of Algerians (some estimates are as high as 45,000) died.

It was in this environment that the Algerian National Liberation Front (FLN) began fighting a war for independence from France in 1954 that lasted until 1962. It was one of the bloodiest wars in the late twentieth century. By its conclusion, an astounding one million people had been killed. The government that emerged from this protracted war of independence soon became a one-party state ruled by the authoritarian leaders Ahmad Ben Bella and Houari Boumédienne. In 1991 the Algerian people attempted to move away from one-party rule to a multiparty democratic system through free elections. The Islamic Salvation Front, an Islamist party, won the first round of elections. The army quickly intervened, canceling the elections and preventing the Islamists from taking control of the government. Law and order broke down. Internal factional war began, pitting Algerian against Algerian and Muslim against Muslim. The Algerian Civil War lasted until 2002.

Finally a political compromise ended the war and paved the path to a representative government. Algeria's current president, Abdelaziz Bouteflika, was elected in 1999. However, his election was marred by accusations of vote rigging, and six of the seven candidates withdrew from the contest. The subsequent 2002 parliamentary elections reflected a general disgust with electoral politics, with only 46 percent of the country voting (including 5 percent who submitted blank ballots as a form of protest). The 2004 presidential elections were considered fairly transparent and free, with Bouteflika being reelected with a huge majority. But the 2007 parliamentary elections resulted in another low turnout, with the ruling coalition winning most of the seats. Algerians are not inspired to vote for a spineless parliament and powerless prime minister, given that unlimited authority is vested in the presidency.

Libya

Libya is an example of another postcolonial North African Muslim state without democracy. The area comprising the modern Libya was under Italian colonial rule starting in 1911. During the 1920s, many Libyans fought Italian colonial forces in an increasingly brutal conflict for independence. After a truce was signed with Italy in 1928, Italian forces arrested and deported tens of thousands in order to quash the nationalist independence movement. After the fall of Axis Italy to the Allies in World War II, the United Nations granted Libya independence. Libya was a constitutional monarchy from 1959 through 1969, when a 1969 military coup overthrew the monarchy and declared Libya a republic.

In reality, the coup's leader, Muammar al-Qaddafi, was and remains Libya's omnipotent leader. All power rests in his hands. Libya's defining years as a state were under repressive colonial occupation. Democratic institutions and ideas were not allowed to grow and mature. Thus, when Libya won independence, it had no democratic roots on which to build a viable and functioning democracy. On the yardstick of democratic opportunity, Libya was at a clear but not insurmountable disadvantage. President Qaddafi's dictatorship could not have been sustained for such a long period if there had been widespread demand for political reform.

Tunisia

Tunisia was the third North African Muslim state under colonial rule as a protectorate of France. As early as 1890, a group of young Tunisians demanded democratic governance. In 1920 the newly formed Doustor (Arabic for "constitution") political party petitioned the French authorities, demanding a constitutional government for Tunisia. The French responded by jailing the leader of the party. Two years later the bey (monarch) demanded a constitutional government. The French responded by surrounding his palace with tanks and forcing him to back down from his position. In 1938 the neo-Doustor party (a splinter of the Doustor party) began organizing disturbances. In 1951 the French allowed a pro-nationalist government to be formed, but when the new government asked for an elected parliament, the French arrested key members of the government. Finally, in 1956, France reluctantly granted Tunisia independence. The country had little experience of democracy, a civil society, or democratic governance. Given this colonial experience, it's not surprising that a half century later Tunisia once again has a system that is rated "not free."

Morocco

Morocco, which has a combined French and Spanish colonial experience, has made strong moves toward democracy in recent years. The first few decades of independence were marked by elections, but almost total dominance of the political sphere was exercised by the king. As events have played out, the king has focused on economic issues in the public sphere. He has made the argument that economic development is a more pressing issue in Morocco than political liberalization. Thus a power-sharing status quo has developed, in which the Parliament focuses on economic and social issues and the king focuses on the rest. The 1992 Constitution and subsequent amendments have increased Parliament's power with such measures as the power to dissolve the government. The most recent 2007 elections were seen as fair and free and marked the third successful parliamentary election in a row. However, the king remains the sovereign in Morocco and firmly controls security and foreign affairs issues.

Egypt

Egypt, although part of North Africa, merits special attention. It is the Arab world's largest and most culturally important nation, and it also experienced colonial exploitation and political intervention similar to the Iranian colonial experience. Starting in 1866, aspects of democratic governance began developing in Egypt. Beginning in the late nineteenth century, Britain began to increase its influence there. It used that influence and power to quash the nascent democracy in the Arab world's strategically most critical state. Although the British never classified Egypt as a colony and named it a protectorate for only a short time, Britain was its de facto ruler for decades through its military dominance and its control over King Farouk. For the British, Farouk was the local instrument of democratic suppression in Egypt, just as in Iran the Americans used the autocratic shah.

The most significant indicator of early and strong Egyptian democratic development can be seen in the Wafd party, led by Saad Zaghlul Pasha and later Mustafa an-Nahhas Pasha. This populist party was extremely popular within Egypt, expanding in power and authority even as World War I consumed the world. Despite its being repeatedly dissolved and elections being rigged against it, the Wafd party kept being reelected again and again, demonstrating the Egyptian people's desire for independence. Thus democratic procedures and institutions were not only evident in Egypt but were actually functioning.

Great Britain's actions against the Wafd party and more widely against Egyptian legislative institutions must be considered a critical element in the undermining of Egyptian democratic growth and the continued autocratic rule of a series of kings and dictators. (Here again, one can only wonder how different the history of the Middle East, the Arab world, and the Muslim world would have been if Egypt, the cultural leader of the Arab world, had been encouraged and allowed to develop a sustained, viable democratic government. It could very well have triggered the domino effect of democracy throughout the Middle East that the United States hoped to achieve with its intervention in Iraq in 2003.)

The Wafd party did not develop out of whole cloth. It was part of Egypt's long history of democratic institutional development. Starting in 1866 with the creation of a Consultative Assembly of Delegates, formed under an elementary constitution called "the Fundamental Law," the people of Egypt attempted to gain political control over their lives. (The notion of "consultation" in Islam, of course, is directly traced to the Holy Book, as discussed in chapter 2.) The clear pattern of Egyptian democratization has been slow and often painful, derailed by Western support for authoritarian regimes that "played ball" with first Paris, then London, and finally Washington. But some forms of democratic institutions have been central to the Egyptian national experience for more than a century.

Soon after the creation of the Consultative Assembly of Delegates, a quranic approach to democracy, the Suez Canal was built by the Suez Canal Company. The company was privately controlled by French investors, with the Egyptian government maintaining a minority share and no authority. In 1875, mired in debt to France and Britain, the Egyptian government sold its 44 percent share in the canal to the British government, making the canal a wholly owned entity of France and Britain. The Egyptian ruler at that time, Khedive Ismail, found himself increasingly in debt to the Anglo-French alliance and abdicated much of his own governmental responsibility to the governments of Britain and France, which gained control over many ministries of the government, including finance and trade.

The critical importance of the Suez Canal to Britain had a negative impact on democratic development in Egypt. The British had a number of important interests at stake in Egypt, but they were interested primarily in keeping the Suez Canal safe, operational, and under their control. As a maritime nation and world power, Britain needed the Suez Canal under its domination for economic, military, and political reasons. Additionally, India, the "jewel" of the British Empire, was accessed by way of the canal. To maintain control over the Indian Subcontinent, Britain needed to maintain control of the Suez Canal and functionally control the government of Egypt. Consequently democracy suffered as it was put on the back burner behind economic and political interests.

In 1879 a charismatic military leader named Colonel Ahmad Urabi, along with some midlevel army officers and small landowners, protested British and French domination. A new, reformist prime minister was appointed by Khedive Ismail. He petitioned for a more equitable debt system, a reassertion of Egyptian sovereignty over all elements of the Egyptian government, and a democratically elected parliament. The response from the colonial powers was swift and direct. Britain and France intervened, forced Khedive Ismail from his throne, and replaced him with his more pliant son, Khedive Tawfiq.

But Tawfiq, lacking popular support, had great difficulty governing. To consolidate his rule, he appointed Mustafa Riyad prime minister. Riyad immediately dissolved the Consultative Assembly. But Colonel Urabi continued to lead a populist movement demanding fundamental social reform. He took to the streets demanding not only the restoration of the Assembly but an increase in its power. Urabi became war minister and led the restored Assembly to pass a Basic Law giving it power over the budget, legislation, and government oversight—in other words, creating a functioning, operative parliament. The British and French, threatened by the real possibility of the creation of an Egyptian democracy and the loss of Anglo-French control, intervened and overthrew Urabi. Urabi fled to Alexandria, where he had strong political support. The people took to the streets in support of Urabi and in support of Egyptian democracy. When the British navy unleashed a brutal barrage of fire against Alexandria, Urabi boldly asserted Egyptian sovereignty and actually declared war against Great Britain, the superpower of the colonial period. Not surprising, Britain—the great land and sea power of the period—had little trouble crushing the Egyptian forces. Then it exiled the democratic populist Colonel Urabi and restored the compliant young King Tawfiq to the throne, where he served as an autocratic dictator for the next decade.

This pattern of intervention in Egypt and British dominance continued into the new century. An unelected Legislative Assembly replaced the elected Consultative Assembly in 1913. When the Ottomans sided with Germany at the outset of World War I, Britain unilaterally declared Egypt a British protectorate and installed a series of compliant puppet rulers on the throne. To constrain the reassertion of democratic development, and specifically to discourage the formation of political parties, the British banned meetings of more than five people and flooded the country with British soldiers.

The declaration of Egypt as a British protectorate and the attempts by Britain to ban free association and free speech caused a significant reaction in Egyptian society. In 1918 a prominent lawyer named Saad Zaghlul petitioned Britain for Egyptian independence. In response to the rejection of his demands, he formed the nationalist Wafd (Arabic for "delegation") party. He organized massive Wafd support that brought down the royalist, pro-British government. In response, Britain exiled Zaghlul and attempted to break up his young political party. But these efforts to contain Egyptian nationalism proved unsuccessful, with the street demonstrations growing increasingly large and widespread.

Against this background of public protest Britain decided to grant Egypt independence. However, London maintained four basic rights that many believed made Egypt into a de facto protectorate: the power to protect communications and control the Suez Canal; the authority to defend Egypt (which was the legal justification for the British occupation of Egypt in World War II); the right to protect foreigners; and the power to determine whether Sudan was to be a separate state or be part of Egypt.

The newly enthroned King Fahd declared Egypt a constitutional monarchy, but he retained the critical power to dissolve Parliament. A secular, democratic constitution was adopted, and parliamentary elections held in 1924 resulted in an overwhelming mandate for the populist, nationalist Wafd party and its leader, Zaghlul, who captured 195 of the 214 seats in Parliament. After the British governor-general was assassinated, King Fahd dismissed Parliament at Britain's request. New elections were held, and despite interference by the British occupation forces and overt rigging by the king's representatives,

Wafd again won convincingly. An almost comical series of interventions, followed by new parliamentary elections, always resulted in the same outcome.

The Wafd party, against all odds, went on to win parliamentary elections again in 1928. Only three months later the king dissolved Parliament on the orders of the British. The Wafd won again in the 1930 parliamentary elections and the 1936 elections. In 1937 the king and Britain managed at last to so tightly control the election machinery that finally the Wafd was prevented from winning an election. But the contest was viewed as fraudulent, rigged, and illegitimate. Lacking popular support, the puppet government that was installed could not function. Seven years later, in the next free election, the Wafd won once again, but King Farouk, under orders from London, dismissed the parliament. It was a comical yet tragic spectacle.

The current Egyptian leader, Hosni Mubarak, took power in 1981, following the assassination of his predecessor, Anwar Sadat. Once in power, Mubarak reinstated emergency law, which had been in effect in Egypt for all but five months since 1967. Initially, the new president seemed to be somewhat reform-minded as he allowed greater press freedoms and freed some political prisoners. He allowed the opposition parties to gain greater numbers against the president's ruling National Democratic Party. But this progress soon stopped as Mubarak reversed course, largely in response to unrest caused by a movement called the Muslim Brotherhood. Press rights were curtailed, opposition leaders arrested, and citizens imprisoned without formal charge.

The 2000 elections were deemed fairer than those in the past, but harassment of voters and manipulation of votes still occurred and Mubarak's NDP continued to rule. In 2005 Mubarak was re-elected for the sixth time, in a plebiscite. Most recently, in the spring of 2007, emergency law was codified in the Constitution. As usual, turnout was low. Although emergency law has now been lifted, its provisions—including detention and arrest powers for police agents in "terrorist" cases—have actually been enshrined in the Constitution. But, like Pakistan's Musharraf, Egypt's Mubarak is viewed by the West as an ally in the war on terrorism. Therefore little pressure has been exerted on him to democratize politically or reform economically.

Egyptian democracy still struggles against a nondemocratic regime that maintains international support despite the fact that it openly rigs elections, victimizes political party leaders and political party organizations, disrupts the working of NGOs (especially those that deal with human rights and political liberty), and allows little press freedom. Once again, an international political strategy trumps democratic values. For the West, once again, democracy is a selectively applied, nonuniversal value, central to foreign policy only when it is convenient and doesn't interfere with other political goals such as the Cold War or the war on terrorism. And it seems that an important result of Mubarak's crackdown on democracy, and in particular his assault on the Muslim Brotherhood, has been a surge in support for religious and political extremism. Turning up the heat on a pressure cooker will eventually make the pressure cooker explode.

Iraq

Another important and tragic example of the failure of the West to promote democratic infrastructure in a critical Islamic state is the poignant case of Iraq. Iraq was awarded to Britain as a mandate following World War I. In Iraq, the British initially tried to emulate their rule of India, even bringing in Indian rulers to govern. This experiment failed, and in 1922 Britain installed the Hashemite King Faisal as ruler. The Hashemite family were Sunni in a heavily Shia population and was generally viewed by the people as foreigners. Indeed, they were, coming from the Arabian Peninsula. Thus the sectarian schism that would tear Iraq apart in a bloody and protracted civil war between the Sunni and Shia populations after the American invasion of Iraq in 2003 was hardly new. And although the Sunni-Shia divide is seven hundred years old, its modern incarnation can very well be traced back to Britain's decision to impose Sunni rule on a Shiite state eighty-five years ago. This minority suppression of majority rights would become the hallmark of the brutal domination by the Sunni Baathists against Iraqi Shiites and Kurds for thirty years at the end of the twentieth century.

In 1925 Britain signed a one-sided oil concession with the Iraqi government, which it controlled, putting most of the profits in the hands of the Turkish Petroleum Company, which in turn was half owned by the British government's Anglo-Persian Oil Company. Although the people of Iraq did elect nationalist parliaments during the time of British rule, the king, acting with the backing of London, dissolved parliaments and called for new elections. The destabilization of nascent democratic institutions in Iraq prevented the emergence of liberal, democratic, and nationalist political leaders. Most observers believe that the intransigence and inflexibility of the British-appointed king resulted in Iraqis' turning to violence to effect political change. Thousands of Iraqis participated in a series of unsuccessful and bloody revolts. Denied the opportunity to govern themselves or affect the quality of their lives through democratic processes, the modus operandi of indigenous Iraqi politics became control of the street. It would manifest itself later in post-independent Iraq in the form of successive military backed coups, and later in post-Saddam Iraq in a bloody and seemingly endless sectarian civil war.

By the end of the British rule, the people of Iraq had been subjected to a succession of British-supported autocratic despots. Following British rule, a series of coups was broken in 1968 when the Baath party, partly led by Saddam Hussein, seized control of the government. Saddam soon controlled the country through manipulation of the security apparatus and the imposition of an iron-fisted rule. When he officially proclaimed himself president in 1979, any possibility for peaceful democratic evolution in Iraq all but ended.

London is not alone in having to take some of the responsibility for the failure of democratic institutions to grow and mature in Iraq both before and after independence. It is quite startling, in light of the two wars fought by the Americans against Saddam's Iraq, in 1990 and 2003, that the U.S. government gave material and significant political, military, and economic support to Saddam during his 1980–1988 war with Iran. The assistance was both covert and overt, and aid supported by senators such as Robert Dole was so pro-Saddam (agricultural subsidies, for example) that the media jokingly called Dole's party "Saddam Hussein's Other Republican Guards." It wouldn't be the first time (or the last, as we shall see in our discussion of Afghanistan) that a nation empowered a Frankenstein monster that turned against it, but it must be one of the more powerful and ironic examples of a policy turned upside down with ongoing historical consequences.

The Western record in Iraq may have led both directly and indirectly to the current state of affairs. The center ruptured with the overthrow of Saddam's Sunni-dominated dictatorial regime, but the American occupiers were not prepared to fill the leadership vacuum. This war of choice, which had been successfully planned on the battlefield but was disastrously ill planned for what would follow, resulted in an extended sectarian battle for control of the country's land and resources. A de facto independent Kurdistan has emerged in the north, and an all but de facto Shiistan in the south. The western part of the country seems to be moving inevitably toward de facto Sunni domination. While the de facto partition of Iraq under a loose federal center would not be an illogical endgame to the current situation, the situation in and around the nation's capital, Baghdad, muddies the waters. Because Baghdad is ethnically mixed, partition cannot become a regional phenomenon. Instead, street-by-street ethnic cleansing is the tragic reality. It is a quagmire for the West and a great and unfolding tragedy for the people of Iraq. The United States finds itself in a colonial war in a postcolonial era, with no way to extricate itself from a mess that it plunged itself into, resulting in chaos for Iraq, the region, and itself. There are no fundamental democratic institutions on which to build an Iraqi democracy. Thus the stated subtext of the American invasion—the creation of a democratic Iraqi state that would become the first domino in the democratization of the Gulf and all of the Middle East—is seemingly unattainable.

Afghanistan

When discussing roads not taken and opportunities missed in the Islamic world, Afghanistan holds a unique place. For a quarter century, Afghanistan has served as the principal venue for the two great international confrontations of the late twentieth and early twenty-first centuries: the Cold War between the West and the Soviet Union and the ongoing terrorist campaign by Islamic extremists against the West. In fact, the two great battles of our era are intrinsically linked on the soil and in the blood of Afghanistan. Decisions made in the 1980s War in Afghanistan, in which the United States financed and armed a local jihad against the Soviet Union, sowed the seeds of the current battle between international terrorists and the civilized world. Indeed, it was the United States, working with the Pakistan Intelligence Service (ISI) to break the backs of the Soviets in Afghanistan in the 1980s, that resulted in the empowerment of the extremist elements of the mujahideen who would, fresh from their role in the destruction of one of the world's superpowers, turn their jihad to the remaining superpower in the early days of the twenty-first century.

The USSR invaded Afghanistan on Christmas Day 1979 in order to prop up the weak, unpopular, leftist government of President Mohammad Najibullah. The government was under assault from a number of Afghan rebel groups, largely configured around ethnicity, religion, and geography.

For the United States, the CIA's covert action against the Soviets was a relatively inexpensive and safe way to hobble the Soviet Union by weighing it down in an unwinnable, intractable quagmire that would drain its economic resources and its national will. No one would claim that analysts within the intelligence agency actually believed that the defeat in Afghanistan would serve as the catalyst for the implosion of the republics of the Soviet Union and the liberation of Eastern and Central Europe. The CIA's covert operation funded, trained, and equipped Afghan rebel groups through the ISI. By routing used Soviet-bloc weapons through Pakistan, the CIA's fingerprints were kept off the transfers, although it was hardly a secret within Afghanistan, Pakistan, or the rest of the world that the United States was deeply involved in the jihad against the Soviets. As the covert operation achieved success, however, there was pressure in Washington—from Congress and the White House—to expand the operation. Soon the operation involved hundreds of millions of dollars a year, military camps in Pakistan where ISI agents and the CIA trained Afghan rebels in modern weaponry, including thousands of advanced Stinger missiles, and the access to sophisticated U.S. spy satellite imagery for the mujahideen leadership. (I will detail the CIA-ISI joint operation in Pakistan and Afghanistan in chapter 4.) The covert mission changed from one of stinging the Soviet Union in Afghanistan to pushing it out of the country entirely. Ultimately, its success would go far beyond that.

The mujahideen who fought the jihad against the Soviets in Afghanistan were not monolithic. On the contrary, the group was a seven-element coalition that brought together diverse elements within Afghani society. They were different in ideology, interpretation of Islam, view of modernity, and acceptance of the West. The glue that held the coalition together was the intense hatred of the Soviet occupiers.

Within the mujahideen coalition there were intractable hard-liners and there were relative moderates. The highest-profile extremist group within the coalition was the Islamic Party, the Hezb-i-Islami-Gulbuddin (HIH). It was led by a religious zealot named Gulbuddin Hekmatyar. He has been described as a talented orator, whose charisma was matched by an authoritarian style characterized by brutality and intimidation. Although the HIH did not initially enjoy significant grassroots support in the country, it quickly became notorious for the fanaticism of its fighters and the almost unspeakable brutality of its treatment of Soviet prisoners.

Hekmatyar was not only radical in his antipathy to the Soviets but was outspokenly and quite openly anti-Western.

Hekmatyar's reputation on the battlefield enamored him to both U.S. and Pakistani intelligence. The CIA and the ISI anointed him their golden boy and showered him with funds, training, and the most modern weaponry. In other words, they empowered him among the other elements of the mujahideen coalition. It was a fateful decision that would not only impact the direction of the war against the Soviets but later affect the rebuilding of the Afghan nation-state, the growth of religious fanaticism within Afghanistan, and the emergence of the Taliban, and ultimately allow the Al Qaeda terrorists safe haven and open operation within Afghanistan.

The decision to empower the extremist factions amongst the Afghan mujahideen caused me to say, in a meeting with President George Herbert Walker Bush in the White House in June 1989, "Mr. President, I fear we have created a Frankenstein that will come back to haunt us."

A second element of the mujahideen coalition was the Islamic Society, the Jamaat-i-Islami (JIA), under the leadership of Burhanuddin Rabbani. The JIA was a militarily powerful party dominated by ethnic Tajiks. Although viewed as fundamentalists, there were Turkmen and Uzbek elements within it that were far more pragmatic and flexible in their attitudes toward the West.

A third fundamentalist force within the mujahideen was the Islamic Union for the Liberation of Afghanistan, Ettihad-i-Islami (IUA). Led by Abdul Rassul Sayyaf, this group was formed in 1982 as an extremist counterbalance to the traditionalists and received support primarily from Saudi Arabia and the Persian Gulf states.

Another strong fundamentalist element within the mujahideen was the Islamic Party, Hezb-e-Islami-Khalis (HIK), headed by Maulvi Mohammad Yunis Khalis. The HIK was predominantly

Pashtun and highly chauvinistic, wanting to return to the interpretation of Islam as practiced in the Middle Ages. The HIK were among the most significant groups militarily on the ground against the Russians.

The moderate faction within the mujahideen had three significant elements. The National Islamic Front of Afghanistan, Mahaz-e-Melli Islami (NIFA), was led by Pir Syed Ahmed Gailani. Its members were Pashtuns who were Sufis, believing their leadership was directly descended from the line of the Holy Prophet. The NIFA was a royalist party that wished to unite Afghanistan under the rule of the deposed King Zahir Shah, who had been overthrown in the military coup of April 1979.

A second element within the moderate faction was the Afghanistan National Liberation Front, Jebh-e-Nejat-i-Melli (ANLF), led by Sebqhatullah Mojadeddi. Its goals were not only to establish an Islamic state under Sharia law but also to guarantee and support individual social freedom. It had the fewest troops in the field and thus did not receive significant support from either the United States or Pakistan, despite the moderation of its views. Mojadeddi did emerge as a compromise leader within the disparate elements of the mujahideen and was selected to lead the mujahideen government in exile in March 1989.

The last faction within the mujahideen coalition was the Islamic Revolutionary Movement, Harakat-e-Inqilab-Islami (IRMA), headed by Mohammad Nabi Mohammadi. It received little recognition from journalists during the jihad and thus remained quite secretive. A relatively small group initially founded in Baluchistan in the early 1970s, it was a nationalist party that received little support or assistance from the CIA and the ISI.

Thus the mujahideen was not an ideological, cohesive monolith. There were significant elements within it that were more open to cooperation and civility with the West, and there were hard-liners. But the hard-liners were supported by General Zia, the military dictator of Pakistan. He shared with them an ideological affiliation with

Jamaat-i-Islami and its related groups. He made them the principal allies of the CIA and the ISI on the ground. It should also be noted that within Zia's ISI there were significant elements that shared values and ideology with the jihadist hard-liners in the mujahideen coalition. The ISI, looking beyond the end of the war, seemed keen on developing close working relations with these elements within the mujahideen whom they would try to empower to rule the new Afghanistan and give Pakistan strategic depth by extending Islamabad's influence northward to counter Kabul's traditional ties with India.

When the USSR decided to pull its forces out of Afghanistan, victory was declared by the United States. Those in favor of the operation had seen the action in the context of a global cold war. At the time, few alarm bells were ringing in Washington regarding the billions of dollars of sophisticated high-tech arms that were now in the hands of hardened Islamic warriors, although there was considerable concern about one specific weapon system—the thousands of missing Stinger antiaircraft shoulder-held missiles that had proved so effective and devastating against Soviet planes and helicopters. (These many thousands of missing Stinger missiles remain, in 2007, a gnawing concern in Washington and in all of the nations of the West. Al Qaeda used them—unsuccessfully—against an Israeli commercial jet in Africa, but their threat to commercial air traffic all over the world remains a significant and potentially catastrophic problem.)

As the post-Soviet situation in Afghanistan deteriorated, with different rebel groups fighting one another, the United States packed its bags and left. Although it had invested billions of dollars in the surrogate war in Afghanistan, there was not a corresponding commitment to rebuilding the country and encouraging the development of political pluralism and free market economics. (U.S. congressman Charlie Wilson would later say that the United States had messed up the "end game.") And with the threat of the Soviet occupiers removed, the diverse mujahideen coalition began to deteriorate into fratricide.

Following the downfall of the Najibullah regime, an extremely tenuous coalition of mujahideen attempted to rule the country until

1996. The central government never truly governed beyond Kabul, with the real power lying in the hands of the local warlords. Starting in 1994, a group called the Taliban rose out of the political madrassas of Pakistan. They promised to end civil war in Afghanistan and restore order. Local Afghan warlords who had fought one another bitterly surrendered to the Taliban without a bullet being fired. Even as they took control of the southern Pashtun belt of Afghanistan, bordering Pakistan's Baluchistan province, the Taliban entered into negotiations with the U.N. special envoy to Afghanistan to create a broad-based government. A treaty on the broad-based government was to be signed by the Taliban on November 6, 1996.

However, with the overthrow of the moderate Pakistan Peoples Party government in Pakistan on the night of November 4, 1996, which had encouraged the Taliban to work with the international community, the Taliban took advantage of the political turmoil in Pakistan. The treaty was not signed on November 6.

Even as power was transferred from moderates to the theocratically inclined Pakistan Muslim League (Nawaz) government through the flawed elections of 1997, the Taliban became more militant. In Pakistan, Prime Minister Nawaz Sharif was praising the Taliban state and promising to emulate it. He too wanted to pass legislation granting authoritarian powers to the chief executive. The opposition feared that Sharif wanted to become *"Amir ul Momineen"* ("guardian of the faithful") like Taliban leader Mullah Omar.

As Islamabad tried moving legislation draped in religious wording, the Taliban became more ruthless. It invited Al Qaeda in, permitting it to raise recruits from the Muslim world and train them in terrorism. It was from the soil of Afghanistan in 1998 that Osama bin Laden would declare war on the West.

I was the Leader of the Opposition in Pakistan's Parliament at the time. Horrified by the open declarations and preparations for war against the West from the soil of our Muslim neighbor and its portents for regional and global security, I called upon the regime of Nawaz Sharif to break its ties with the Taliban. I was convinced that

Islamabad could influence the Taliban into stopping its support for Al Qaeda. Afghanistan is a landlocked country that must traverse Pakistan to the Arabian Sea on the shores of Pakistan for its trade.

However, the government of Prime Minister Sharif refused to heed my words of warning. Intoxicated with the thought of giving "strategic depth" to Pakistan's northern frontiers through a pliant Taliban government in Kabul, they mocked my warning in Parliament that instead of giving strategic depth to Pakistan, the Taliban would become "a strategic threat" to the welfare of Pakistan. Sadly that's just how it turned out after 9/11. Driven from Afghanistan, the Taliban went into the tribal areas of Pakistan to regroup and reassert themselves.

The Taliban's rule in Afghanistan was ruthless. It often brutally imposed their rigid, puritanical views on Afghan society. It forced men to wear beards. They made it impossible for girls and women to attend school. They destroyed any vestiges of other religions or other cultures. Only three states ever recognized the Taliban as the legitimate government of Afghanistan, and the regime enjoyed little economic success.

Following the 1998 attacks on two U.S. embassies in Africa, the United Nations demanded that the Taliban extradite Osama bin Laden, who was an official guest of the Afghan government. When Mullah Omar refused, the United Nations imposed harsh economic sanctions, further isolating the backward regime. Bin Laden and his September 11, 2001, attacks on the United States proved to be the Taliban's downfall. After Afghanistan once again refused to turn the tall Saudi over, the United States began major military actions against Afghanistan in October 2001. Mullah Omar and the Taliban leaders surrendered their forces in December and retreated to the Tora Bora mountains. They are thought to have escaped from Tora Bora into the tribal areas of Pakistan.

The U.S. campaign in Afghanistan was widely viewed as a success, with plans for elections being made and a constitution being written. The Taliban, however, had spent the winter of 2002 regrouping in the south of Afghanistan and in the border areas with Pakistan. Starting in 2003 it began using guerrilla tactics in a renewed effort against U.S. and NATO forces.

Afghanistan has held successful elections and has written a constitution. However, the Taliban has now clearly reconstituted in safe havens in the tribal areas of Pakistan in territory functionally ceded over to them by the Musharraf military regime. A resurgence—magnified since Pakistan's 2006 military pullout from north Waziristan—has made Afghanistan dangerous once more and negatively impacted the integrity and unity of Pakistan.

Though the United States supported the mujahideen financially, it was always held at arm's length from the mujahideen by the ISI. There was much appreciation of U.S weapons among the mujahideen but not much love for the United States. By pulling out of Afghanistan in the early 1990s, the United States lost all control over, influence on, and intelligence about the radical groups that it had financed. Suddenly there were thousands of U.S.-trained and -financed radical fighters left out in the cold. There is also little mystery as to why the ISI has been slow to cut connections with Taliban and Al Qaeda groups that it has fought alongside for years. Many of the officials in the ISI during the dictatorship of the 1980s were tied to the religious party known as Jamaat-i-Islami and to the works of its founder, Maulana Maudoodi. Moderates had been purged by the dictatorship, fearing their sympathies for the executed reformist and the democratically elected prime minister Zulfikar Ali Bhutto. In fact, a "beard allowance" was introduced by the military dictator into the armed forces to reward those who kept their beards. Beards were seen by some as a sign of religious obedience.

Some of the intelligence officials who had fought the Soviet Union through their proxies in the Afghan mujahideen had brainwashed themselves into believing that they themselves had actually defeated the Soviet Union. They believed that having defeated one superpower, they could defeat another. I was surprised at the selective amnesia they had, for it had been the international community that used the

then ISI as their surrogates to fight the Soviet occupation of Afghanistan. The international community had provided satellite images,
weapons, intelligence, and political and diplomatic support, as well as
plenty of cash.

Having since retired from the military, these ISI officials have tried
to maintain their grip on power by destabilizing democracy and
politically reengineering Pakistan's political structure by creating a
National Accountability Bureau (NAB). Ostensibly the job of the
NAB is to root out corruption. However, its laws have been used to
root out moderate political opposition. Given the focus on crushing
political opposition, it was not surprising when Transparency International—the Berlin-based NGO that is considered the key global
civil society organization leading the fight against corruption—in its
2006 report, found that corruption under General Musharraf was
higher than under his civilian predecessors.

With the destabilization of democracy in 1996, extremism has
risen in Pakistan, growing with every successive government. Under
President Farooq Leghari's interim government, the Taliban reneged
on the U.N. initiative for a broad-based government that my government had backed and was to be signed on November 6, 1996. Under
Prime Minister Nawaz Sharif, a constitutional amendment was
moved that aimed to incorporate Taliban principles into Pakistani
law. However, before it could be bulldozed through the Senate,
Mr. Sharif fell out with the military and General Musharraf seized
power in a military coup in October 1999. Although General Musharraf said he admired the secular Turkish reformer Mustafa Kemal
Atatürk and was pictured with two dogs in his arms like Atatürk,
the growth of militias and militants in Pakistan has been stupen-
dous. They now control large swathes of the tribal areas and have
established military depots, disguising them under the name of
madrassas. The Red Mosque mutiny in Islamabad in July 2007 was
an eruption of the threat that has spread its tentacles into the cities of
Pakistan.

Even though the regime finally took on the militants of the Red

Mosque with a huge loss of one hundred lives, it surreptitiously allowed the militants back through a court order. The court ordered that the Red Mosque Madrassa Complex be handed over to the militants. This, despite the madrassa complex having been built illegally without building permission by the Capital Development Authority and on illegally occupied government land belonging to the Education Ministry for the building of a children's library.

To some extremists in Pakistan who battle against moderation and modernity, the Taliban are ideological partners and brothers in battle against the West. A Taliban-ruled Afghanistan also gives the Pakistani military the "strategic depth" that it has long desired. The decision of the United States to abandon Afghanistan in 1989 instead of rebuilding it continues to ripple not only across the Central Asian landscape but across the entire world.

Comoros

The little archipelago of Comoros is an example of one of the less significant venues of Western antidemocracy intervention. The island became a French colony in 1886 and declared independence in 1975. The country developed as a socialist, secular democracy until a French-supported coup, backed by European mercenary soldiers, toppled the government and returned Comoros to one-party, dictatorial rule for the next generation.

Lebanon

Lebanon is another important example. It came under French control at the conclusion of World War I. During World War II the Free French government helped proclaim independence for the Lebanese but did not, in fact, grant independence. When the Lebanese held elections in 1943, the French arrested the newly elected government. The French finally granted independence to Lebanon in 1946, but since then the development of democratic infrastructure in Lebanon has proceeded only in spurts, with Western interest and support correlating with external factors. The West's sudden reinterest in Leba-

nese democracy as a check on growing Syrian influence was disrupted
by the recent electoral inroads of Hezbollah. With Hezbollah a growing threat to Lebanese national integration and Western interests in
containing both Syria and Iran, U.S. and French support for democracy in Lebanon has diminished precipitously.

Pakistan and Bangladesh

There have, of course, been qualified success stories of democratic
development in the Muslim world. Certainly my homeland—the
Islamic Republic of Pakistan—has experienced both thriving competitive democracy and brutal dictatorship. Although much attention
has been given to the brutality of the Zia dictatorship in the 1980s
and the illegitimacy of the Musharraf dictatorship over the last
eight years since his coup against an elected civilian government,
democratic institutions—political parties, NGOs, independent
media—have been sustained, developed, and strengthened. These
key democratic institutions have managed to survive even the most
violent and targeted assaults by authoritarian regimes. Indeed, frontal
attacks have often triggered popular responses that have rekindled
the call for democratic change. Certainly the sacking by General
Musharraf of the chief justice of the Supreme Court led to a public
protest in Pakistan. People in Pakistan remain optimistic that democracy can be sustained, and I have devoted my life to that goal.

Democracy has also seen phases of success and electoral competition in Bangladesh. Although the country is currently under military
rule, the periods of democratic governance should not be dismissed
and hopefully have built the seeds of a democratic resurgence in the
future. Democratic development in Bangladesh may have been arrested, but there is a democratic infrastructure waiting to be reconstituted. It remains an interesting, if not confusing, example of a Muslim
country attempting democracy. It is a partial success story, although
currently ruled by a military junta that has imprisoned the leaders of
the two largest parties, both of whom are former prime ministers.

Partition of the Indian Subcontinent in 1947 resulted in a bifurcated Pakistan. West Pakistan (present-day Pakistan) was located on the western border of India, while East Pakistan (present-day Bangladesh) was located on the eastern side of India.

The 1956 Constitution attempted to bridge this geographical chasm of a physically divided Pakistan by giving East and West Pakistan equal representation in the National Assembly. In 1958 the tone changed somewhat when the East Pakistani Provincial Assembly voted for autonomy in all fields except foreign affairs, defense, and currency. Sadly, the issues between East and West Pakistan never had a chance to be resolved directly by those in both sections of my country who wanted to bridge the divisions that had developed. This was because in 1958 General Ayub Khan dissolved the Constitution and dismissed the National Assembly, declaring martial law. The political process was abruptly frozen, and the chance for peaceful accommodation of competing values within the system became extremely problematic.

When unrest forced General Ayub Khan from office in 1969, he replaced himself with General Yahya Khan, who was to supervise Pakistan in its transition back to democracy and national reconciliation. The 1970 national elections proved to be a watershed moment. The Awami League won almost all the National Assembly seats in East Pakistan after its main competitor, Maulana Bashani's party, boycotted the elections in the eastern wing. It gained a majority of seats in the Pakistani National Assembly, although it did not win a single seat in any of the federating units of West Pakistan. The Pakistan Peoples Party of my father, Zulfikar Ali Bhutto, won a majority of the seats in West Pakistan but did not win any seats in the eastern wing. General Yahya's elections simultaneously created both a Constituent Assembly and a Legislative Assembly. The Constituent Assembly was to frame a constitution, while the Legislative Assembly was to govern, although the members of both assemblies were the same.

Although Sheikh Mujib, the leader of the majority party, had the sole right to form the government (even though he had no representation from the western wing of Pakistan), he did not have the right to impose a constitution on all the people and federating units of Pakistan. He did not have one member representing the remaining four federating units and the people elected on his ticket. General Yahya ordered that the constitution must be completed within 120 days or the Assembly would be dismissed. Meanwhile, Sheikh Mujib had formed a militia that started dismantling the state from within. His Mukti Bahini militants took on police duties and provoked the army. It seemed a matter of time before Sheikh Mujib would announce a unilateral declaration of independence. Separately Khan and my father attempted to negotiate a realistic compromise with Mujib, but the Awami League wanted all or nothing.

My father was in Dacca the night General Yahya Khan ordered a military crackdown. General Yahya Khan then suspended the formation of the National Assembly, and Mujib and his Awami League responded with a general strike in East Pakistan.

Responding to this military crackdown, a Bengali officer by the name of Major Ziaur Rahman declared independence for East Pakistan. India, taking advantage of the situation and always fearing a strong, united Pakistan, allowed the leaders of the Awami League into India to form a government in exile. Relations between India and Pakistan, always bad, deteriorated as millions of Bengali refugees poured across the border into India. In early December, Indian troops attacked Pakistan. The Pakistani Army was dramatically understaffed, underfinanced, and undersupplied compared to the huge Indian Army, and though the fighting was fierce and difficult, Pakistan could not sustain the battle against the Indians. On December 16, 1971, the Pakistani Army surrendered at Dacca Racecourse Ground.

Sheikh Mujibur Rahman, the leader of Bangladeshi independence, was elected prime minister of the new country of Bangladesh. In the midst of an economic crisis that swept the new country in 1975, Mujibur Rahman assumed extraordinary extraconstitutional powers. He was killed in a coup later that year, and eventually General Ziaur Rahman was able to take over control of the country. He was officially elected president in a 1977 referendum that legitimized his rule. In 1981 Rahman, like his immediate predecessor, was felled by an assassin. Elections followed that year, but shortly after the new elected government was installed, General Hussein Mohammad Ershad took over control of the country in a military coup.

Democracy was restored to Bangladesh when General Ershad stepped down in 1990. During the next sixteen years Khaleda Zia (widow of the assassinated General Ziaur Rahman) and Sheikh Hasina (daughter of the assassinated Sheikh Mujibur Rahman) alternated terms as prime minister in robustly competitive national elections and badly polarized parliaments. Charges of corruption were hurled by both at each other. Since the end of Khaleda Zia's last term there have been no elections in Bangladesh and a caretaker government has been installed by the military. As of now, both Khaleda Zia and Sheikh Hasina are in jail and a military-supported junta rules the nation of Bangladesh. One of the obvious goals of the military regime is to undermine the two largest parties in the country by creating new centers of leadership and destroying the existing political infrastructure. Since both national parties in Bangladesh have formidable grassroots support, the military junta's attempt to break up the party structure will not be easily accomplished.

Bangladesh is one of the most densely populated countries in the world, is prone to natural disasters, and has a challenged, developing economy. Democratic governance is inherently difficult in the developing world, but the very young and fragile nation of Bangladesh has been particularly handicapped. After almost two decades of democratic elections and governance in Bangladesh we see the nation's two most prominent leaders in jail and a caretaker government controlled by the military ruling the nation. However, a population that has developed the expectation of civilian control and has exercised the privilege of democratic choice in open and hotly contested competitive elections cannot long be denied the ability to be directly involved in the selection of its national leaders and national Parliament. Democratic development in Bangladesh may be arrested, but there is sufficient infrastructure in place to believe that democracy in that nation can once again be restored.

Jordan

The Kingdom of Jordan has, for most of its existence, enjoyed a warm and close relationship with the West. The territory of modern Jordan was part of the Palestine Mandate, which was a British possession after World War I. Although the modern state of Jordan broke off from Palestine in 1921, Jordan continued to be inextricably tied to Palestinian affairs. Transjordan (as Jordan was known before independence) was a monarchy under King Abdullah in 1921. In 1929 a Legislative Council was set up, and in 1939 this council became King Abdullah's cabinet. This democratic growth came through generally peaceful cooperation and consultation with the British colonial lords.

The 1948 Arab-Israeli War had a huge impact on Jordan both emotionally in terms of the military defeat and demographically because of the absorption of 500,000 Palestinian refugees to be added to the Jordanian population, which at that time was only 340,000 citizens. Any Jordanian democratic movement and developing democratic infrastructure would now have to include the West Bank, which had been annexed by Jordan during the war.

Jordan was technically a constitutional monarchy, but the king retained extraordinary powers not consistent with parliamentary democracy, most notably the power to dismiss Parliament. Since Jordan's population was majority Palestinian, this power of sacking governments was a check on Palestinian expansion and takeover of the Hashemite Kingdom. In 1952 the young and progressive King Hussein ascended the throne in what would be a half century of rule.

Before the creation of the Palestinian Authority and Jordan's forfeiture of claims to the West Bank, the strongest challenge to the Jordanian monarchy's power had consistently come from the Palestine Liberation Organization and other Palestinian groups, which used the territory of Jordan to conduct cross-border attacks against Israel.

King Hussein decided to actively participate in the 1967 war in support of Egypt and Syria. It was a decision that had profound consequences. Jordan was thus part of the Arab military defeat and lost political control of the West Bank and East Jerusalem.

After the war Jordan was forced to absorb still more hundreds of thousands of Palestinian refugees who fled the West Bank and the Israeli occupation.

This demographic shift further exacerbated the Palestinian-Jordanian tension. In 1970 guerrilla groups known as fedayeen (Arabic for "one who sacrifices himself") began openly challenging the Hussein regime. In September 1970, in response to a series of airliner hijackings, King Hussein began openly confronting the fedayeen for control of Jordan in a military operation that came to be referred to as "Black September." By 1971 Hussein had won the internal war and expelled the political and military wings of the Palestine Liberation Organization from Jordan. In recent years Jordanian democracy has been challenged—some would say cynically manipulated. In 1989, under King Hussein, Jordan began a reform program to restore some elements of democracy that had previously been disrupted. The Palestinian uprising against Israel (the intifada), along with mounting financial and economic trouble in the kingdom, prompted the king to take action to solidify his hold on power. He called for full parliamentary elections, and the subsequently elected National Assembly legalized political parties and eased press restrictions. The late 1990s brought a reversal of this trend, as domestic discontent over peace with Israel and continuing economic strain prompted King Hussein to once again restrict the press and modify election laws. These actions resulted in a boycott of the 1997 parliamentary elections by the Jordanian political opposition, and thus the pro-regime parties swept the seats in the elected Parliament. In 1999 King Hussein died after a long battle with cancer, and his young, Western-educated son Abdullah II replaced him on Jordan's throne.

An enlightened and articulate young man, the newly crowned King Abdullah promised major economic and political reforms. But the elections scheduled for 2001 were postponed because of the turmoil of the second Palestinian intifada and the regional political situation. In 2003 elections were finally held, conforming to international standards. These elections were generally considered to be transparent, free, and fair, with a number of parties aggressively competing. This was a very important step forward for Jordan. But although this electoral process was an important milestone in Jordan's road to democracy, the gap between democratic elections and democratic governance remains wide. Parliament is representative, but it has limited authority and limited legislative power and responsibility. King Abdullah's government still remains firmly in control of Jordan without formal accountability to the people. Abdullah may well be enlightened, but his regime cannot yet be classified as democratic.

Palestine

Palestine is an interesting case study of the radicalization of a moderate polity as an outgrowth of political suppression and economic deprivation. It is also an example of the conundrum and dilemma of the election of an extremist-dominated parliament through free and fair elections.

Palestine has had a uniquely tumultuous history. In the past century Palestine has been under Ottoman rule, British rule, Egyptian-Jordanian rule, and Israeli rule.

For the last century, Palestinian history has been determined by outsiders. Palestinians have not had a state of their own, and the two small parcels of land that they are allowed to administer are not geographically contiguous. (At various points in the peace process between Israel and the Palestinians, especially after the active intervention of the Clinton administration at the end of the Clinton presidency, various proposals for a physical linkage between the two parts of the Palestinian state through a land corridor cutting through the Israeli Negev desert have been discussed and agreed to. But since the total peace package was never adopted, the plans for a land bridge between Gaza and the West Bank have never been implemented.)

Since 1967 the land that was formerly Palestine has been divided into three major areas. The majority of the land is part of the state of Israel, while two smaller portions of land—the West Bank and the Gaza Strip—are self-governing, semiautonomous Palestinian territories.

After the First World War, Palestine became a British protectorate. During the period leading up to the Second World War, the Palestinian land was promised by the British to both the Jews and to the Arabs. In the aftermath of the war, the United Nations partitioned the land into an area for Jews and an area for Arabs. (It was the second religious partition of that period, coming at approximately the same time as the religious partition of the South Asian Subcontinent into the predominantly Hindu state of India and the predominantly Muslim state of Pakistan.) The Jews accepted the United Nations' plan, but the Palestinians and their Arab neighbors strongly rejected it. The subsequent 1948 Arab-Israeli War resulted in a victory for the newly independent Jewish state and left the Palestinians with the Gaza Strip and the West Bank. The Gaza Strip was annexed and occupied by Egypt, and the West Bank by Jordan. The 1967 Arab-Israeli War was another victory for the Israelis, who captured both the West Bank and Gaza and thus greatly geographically reduced the size of the Palestinian territories. The whole of the city of Jerusalem was also taken by the Israelis during this conflict.

Since the 1967 conflict there have been another full-scale war (1973) between the Arabs and Israelis, various peace accords (Camp David I, Israel-Jordan Treaty of Peace), peace plans (Oslo, Camp David II, the Road Map), Israeli land invasion of Lebanon, and later an air war against Hezbollah in Lebanon and two Palestinian intifadas. There has not been significant progress, however, on the fundamental issue of the Arab-Israeli dispute, which centers on the national aspirations of the Palestinian people. Although a Palestinian Authority has gained limited control over parts of the West Bank and all of Gaza, the peace process is stalled.

Electoral politics have had a limited history in Palestine. During all of its modern history, Palestine has been under one foreign rule or another. There have, however, been elections to determine the government of the Palestinian Authority, which has some autonomous control over parts of the occupied territories, although Israeli security forces retain fundamental control in the West Bank and oversight and occasional intervention in Gaza. Prior to his death in 2004, President Yassir Arafat was the elected leader of the Palestine Liberation Organization and had been the de facto and later de jure leader of the Palestinian people for well over three decades. After his death, elections were held in 2005, and the moderate Mahmoud Abbas (from Arafat's Fatah party) was elected president of the Palestinian Authority—the ruling body of Palestine. During the first post-Arafat elections for the Palestinian Authority, the more extreme and religiously fundamental Hamas party boycotted the elections.

Hamas is a Palestinian organization whose formal charter provides for the use of violence to promote and achieve the goal of the creation of a Palestinian state. Israel and most major Western nations have labeled Hamas a terrorist organization. A major turning point in Palestine's transition to democracy occurred in January 2006. It is significant that despite the political difficulties of living under foreign occupation and having land that is physically separated, the Palestinian Authority did indeed manage to hold democratic, fair elections in 2006. These elections were conducted with the active participation of education and training teams from Europe and the United States. Former U.S. president Jimmy Carter led a large United States joint delegation from his own Carter Center and the National Endowment for Democracy's National Democratic Institute to monitor the results. The Carter delegation, universally respected for its work around the world, certified the elections as free and fair. What emerged for the West was the true paradox of democracy. After years of occupation, violation of political rights, and lack of progress on improving the economic and social quality of life, a desperate people had turned away from the traditional political structures and norms.

Fatah, viewed as incapable of improving the political and economic conditions, was voted out of office. Hamas gained a clear majority of the Palestinian Authority Parliament in Ramallah. It emerged with a resounding victory, winning a plurality of the votes among the combined Gaza and West Bank polity.

The West had pushed the rhetoric of democracy and the practice of electoral democracy. Yet when the results unexpectedly empowered what many consider to be an extremist party, the election was denounced. Indeed, the Hamas victory in Palestine has been used by some circles in the West to caution about the "dangers of democracy" in other places around the world, including, most painfully for me, for the people of Pakistan.

The magnitude of the Hamas victory was exaggerated by strategic errors by Fatah. In a first-past-the-post election, Fatah often fielded several candidates who ran against one another, while Hamas judiciously selected one candidate to carry its banner. Often the Hamas candidate emerged with a plurality despite the fact that altogether, the votes for the various Fatah candidates constituted a majority. The Hamas victory over Fatah was especially crushing in Gaza. Both the European Union and the United States, which continue to characterize Hamas as a terrorist organization, declared that unless certain conditions were met, including the recognition of Israel and the rejection of the use of violence to accomplish its political ends, economic assistance would be suspended. Hamas would not agree to these stipulations, which were inconsistent with its charter and political strategy. It believes that military action in support of the political goal of liberating occupied territory is not, in fact, terrorism. Although European facilitators attempted to mediate and moderate these points, Hamas could not accept Europe and the United States' conditions. Aid from abroad was frozen, as was Israeli transfer of Palestinian customs funds. Palestine's economy was soon crippled.

Following this economic meltdown, the Palestinian Authority could not pay its employees, who began revolting. The economic situation mirrored (and exacerbated) the political situation. With trade with Israel suspended and the border closed to Gazans who previously crossed into Israel to work, unemployment in Gaza has risen to a crippling 60 percent. Since the Hamas government takeover, Fatah and Hamas have been unable to work together in any meaningful way, and their relationship has deteriorated into an all-out war of words and bullets. In March 2007, it appeared as if the internal conflict might end after Saudi Arabian King Abdullah intervened and negotiated a power-sharing agreement between the two parties. But the divisions between Hamas and Fatah proved to be insurmountable and unbridgeable, even with Saudi mediation. The violence came to a head in June 2007, when President Mahmoud Abbas dissolved Parliament. Hamas responded with military action against Fatah security forces in Gaza. A bloody civil war raged for several days before Fatah was driven from Gaza and the territory was under the complete control of Hamas. The Hamas green flag was raised over the land, and Fatah police and security forces were disbanded.

Now there are two separate, semiautonomous Palestinian states. Hamas controls Gaza, while Fatah maintains control over most of the West Bank. Economically and socially, the standard of living and the quality of education, housing, and health in the West Bank far exceed what exists in Gaza. Although it was the intent of Israel, Europe, and the United States to demonstrate to the people of Gaza the consequences of their alignment with Hamas, the strategy may have backfired. It appears that Hamas has yet to be blamed by the local population for the deterioration in living conditions in the territory. There has been no apparent surge in support for the moderate Fatah. On the contrary, it seems that the deterioration of the living standards in Gaza has served to further radicalize the people of the territory. Additionally, with international support suspended, many NGOs that had serviced the territory have ceased to operate. Political and human rights in Gaza have thus witnessed a corresponding decline. There are few remnants of political pluralism and press freedom. In Gaza, democracy has remained an electoral process without being translated into democratic governance.

Democratic governance cannot come out of nowhere; it takes years of democratic institution building. Over the years, the Palestinians have experienced a fundamental abuse of their basic civil and human rights, occupation by foreign military powers, internal and external violence, two massive uprisings, personal and national humiliation, and economic depression. When democracy was suddenly thrust in front of them, most Palestinians chose a radical group that promised them freedom and dignity through violence. Impatient and discouraged, they voted for revolution instead of evolution.

·

The Palestine pattern suggests that it may be a better electoral strategy for moderate regimes to be, at an early stage, inclusive with Islamist parties, encouraging them to participate in the traditional political process. Exclusion seems to fuel radicalization. And, as the comparative democratic scholar Carl Gershman has noted, attempts at inclusion may very well prove to be pragmatic and prescient.

In Jordan, for example, the participation with occasional interruptions of the Islamic Action Front, the party of the Muslim Brotherhood, in elections to the lower house is seen by some analysts as a key reason Jordan has remained stable despite the deteriorating conditions in nearby Iraq and Palestine. In Yemen, the Islamist Islah party participated last September on the losing side of the freest presidential election ever held in an Arab country, an inclusive process seen by President Ali Abdullah Saleh as the best way to keep his fractious country together and to get badly needed foreign aid. In Morocco and Kuwait, Islamists are now represented in the Parliament and will be part of a developing debate over whether and how these traditional autocratic monarchies might evolve into constitutional monarchies on the European model. Turkey is not an Arab country, but the ruling AK party, which evolved out of the Islamist movement and won elections in July 2007 on a program of liberal economic reform and integration into Europe, offers a democratic model for the Arab and the larger Muslim world.

Turkey

One of the true success stories of democratic governance in the Muslim world is the case of Turkey. Following the breakup of the Ottoman Empire, Mustafa Kemal Atatürk fought for and founded the modern Turkish state. Atatürk envisioned Turkey as a secular, nationalist state. As part of his immense reform package he introduced the Latin alphabet to the Turkish language, and the Muslim call to prayer in Turkey was required to be in Turkish instead of Arabic. These changes, at once symbolic and substantive, as well as many others, were deliberate attempts to instill a sense of Turkish, secular nationalism in the people.

The 1924 Constitution provided for a unicameral legislature with a president as head of state. When President Atatürk died in 1938, there was a peaceful, democratic transition of power to İsmet İnönü, testifying to the strength of the system Atatürk had established.

The nation's second political party, the Democratic Party (DP), successfully took power from Atatürk's party, the Republican People's Party (CHP), in 1950. A peaceful transition of power from the incumbent party to the opposition party is perhaps the most significant benchmark of a successful transition to democracy. However, the DP imposed restrictions on civil liberties in 1957, which resulted in violence and eventually the imposition of martial law by the DP. This led the chief of the general staff, General Cernal Gürsel, to stage a coup and overthrow the government. A new constitution was then written and overwhelmingly ratified with 90 percent support by the people of Turkey in a referendum that generated a very high turnout. Elections and a return to civilian rule occurred shortly thereafter, in 1961.

The next challenge to democracy took place in 1971. Süleyman Demirel's government came under criticism when it accepted radical rightist parties into its coalition. The chief of the general staff sent Demirel a memo stating that Demirel should install a "strong and credible" government. Demirel resigned, and a series of caretaker cabinets ruled Turkey until the 1973 elections. This process has been referred to as the "coup by memorandum."

In the 1970s Turkey experienced a substantial amount of political violence, and some governments felt compelled to institute martial law in areas of the country. This violence and unrest peaked in 1980, when a large Islamist rally disrespected the flag and national anthem, two strong symbols of Atatürk's secular vision for the newly independent Turkish nation. A military coup followed. This coup, more repressive then the previous two, resulted in approximately 30,000 people being arrested, many without charge. In July 1982 a new constitution was written, giving the president more power. In November, elections were held and civilian rule was once again restored.

The most recent challenge to democratic rule has been the government of the Justice and Development Party (AKP), headed by Recep Tayyip Erdoğan. Many believed that the party had an underlying Islamist agenda, a charge that Erdoğan strongly denounced and rejected. In 2007 AKP put Erdoğan's name forward as a presidential candidate, but he eventually stood down due to massive protests. The secular opposition, including the powerful military, believed that Erdoğan's presidency would undermine the secular nature of Turkey and turn Turkey in the direction of fundamentalist Muslim states. Some even suggested he would apply Islamic law to the nation, undermining its constitution according to the concept that Atatürk had envisioned. Eventually, after foreboding remarks by the Turkish Army that seemed to threaten a coup if an Islamist candidate were elected president, the Constitution was changed to provide for the direct election, as opposed to appointment, of the president. Following a thumping victory by the AKP in the July 2007 parliamentary elections, Abdullah Gül was elected president directly by the Turkish people in August 2007.

A quick glance at Turkey's past might give one the impression that

Turkey's democracy has been unsuccessful, given the history of coups and the active intervention, from time to time, of the military. This is not necessarily the case. Turkey began as a modern state less than ninety years ago. Its democratic progress has been remarkable, despite great pressure from external and internal stress. Recent events have proven the fundamental strength and vitality of Turkish democracy. When challenged, the AKP government did not resort to extrademocratic applications of power but rather fell back on the processes of democracy by organizing free and fair early elections. In other places and at other times, some extraconstitutional attempts to coalesce power might have been expected. The fact that the ruling party chose a path that actually extended the popular legitimacy of government not only demonstrated the strength of the AKP, following its subsequent electoral victory, but the inherent strength of the developing political institutions of the Turkish nation.

Indonesia

In the pantheon of Muslim countries, Indonesia, the largest Muslim nation on Earth, has also emerged as an example of one of the highest degrees of success in democratic governance. Although Freedom House characterizes Indonesia as one of the Muslim world's only truly free nations, with both political and civil liberty, the young Indonesian democratic system has not yet been sustained to the point that we can be sure the democratic experiment will ultimately succeed. But at this point, Indonesia is a thriving and viable Muslim democracy. Prior to independence, Indonesia was a Dutch colony known as the Dutch East Indies. Starting in 1918, Indonesians could be elected to the Volksraad (the People's Council), which was the colonial governing body. The members could not yet vote, but Indonesian nationalism was beginning to emerge. In 1927 a young nationalist activist by the name of Sukarno helped found the Indonesian Nationalist Party (PMI). Sukarno was subsequently arrested by the Dutch, who, rather typically for an imperialist power, feared his ability to unite the people. The Dutch wanted to keep their colony stable and peaceful in order to preserve their very significant economic interests in the colony. Indonesia was rich in oil and rubber, both of which were playing a vital role in the new automobile-driven world. In the Dutch East Indies, economics and governance were closely tied together. Indeed, the Indonesian colonial governor-general was also the director of the primary oil company in the archipelago, Royal Dutch.

The Japanese occupation during World War II acted as a catalyst for strengthening the concepts of nationalism and individual liberty in Indonesian life. The brutality of the Japanese occupation was a defining moment in the creation of a politically viable Indonesian national identity. When the war was finally over and the Japanese left, the Dutch attempted to reassert their economic and political control of their colony. But the international democratic community rose up against this reimposition of colonialism in the post–World War II era, especially in light of the brutality with which the Japanese treated the people of Indonesia during the years of their occupation. The United States, in an uncharacteristically bold move against an ally, threatened to cut off all financial assistance to the Netherlands, under the impressive Marshall Plan that was being implemented to rebuild the economies of war-torn Europe, if the Dutch attempted to reclaim Indonesia as a Dutch colony. With the support of the international community, nationalism on the archipelago blossomed and matured, and the intense fervor that swept the islands crested in the creation of a new nation, the independent state that was named Indonesia. This historic birth of the largest Muslim nation on earth occurred in 1949, at the very outset of the Cold War between the West and the Communist world.

After independence the new government under Sukarno struggled. Under the Dutch, most of the administration of the country had been run by apparatchiks imported from The Hague; the Indonesians were not trained to take over their country. Thus the administration of the country had to be learned with no preparation and without any institutional support. Not unsurprising, due largely to poor management and a lack of technology, the economy performed poorly and unrest grew. Sukarno declared martial law in 1957 and in 1959 declared the Constitution void, which began a new period of so-called guided democracy. Essentially this meant that most power would be concentrated in the rapidly expanding presidency.

This would all change dramatically in 1965. In September of that year, six high-ranking generals were assassinated by sympathizers and members of the Indonesian Communist Party (PKI). The PKI was closely aligned with President Sukarno. In the following months, General Suharto suppressed the initial coup and then went on to destroy the PKI and its followers. This purge spread to the countryside, and hundreds of thousands of Communists and ethnic Chinese were killed. By 1967 it had become clear that Suharto had consolidated most of the power in the state, and Sukarno had no other choice than to sign a document transferring power and authority to Suharto.

During Suharto's thirty-one-year rule, the Indonesian economy boomed and the standard of living dramatically improved. But despite the economic progress, there was no political liberalization in Suharto's iron rule. Democratic political institution building was not allowed to develop. As Suharto's rule entered the 1990s and grew increasingly authoritarian and repressive, Indonesians became increasingly uncomfortable with the rule of the president and the direction of their nation. The 1997 economic collapse proved to be the tipping point, as rioters and protestors took to the street to demand an end to Suharto's rule. With the popular groundswell sweeping the nation, Suharto realized that his coercive base of power would no longer be enough to keep him as president. To avoid a bloodbath that he would most likely have lost anyway, he resigned in 1998.

Following Suharto's resignation, Indonesia entered a period of democratic blossoming. Numerous reforms to the political system took place, including a new direct election for the presidency. The concept of the supremacy of a popularly elected representative body over a president was institutionalized into law. Furthermore, the new procedures asserted civilian control over a military that had a long record of political intervention. Indonesia became a model for the entire Muslim world of what a true democracy could be.

Indonesia is proving that despite the historical chains of colonialism, a Muslim country can create a successful democracy. In Indonesia, contrary to the conventional wisdom that asserts that Islam is inconsistent with democracy, democracy and Islam have flourished and developed together, one getting strength from the other.

Senegal and Mali

A further look at the rest of the Muslim countries in the world is quite revealing. There are unfortunately only a few clear democratic success stories. Senegal and Mali both stand out as postcolonial Muslim states that have made great achievements in the field of democracy. President Abdou Diouf of Senegal increased democratic participation by holding the first elections in 1983 and widening participation. When in 2000 he lost an election to a candidate from another party, he peacefully let go of power. This was the first peaceful interparty transfer of power in Senegal. Elections were considered free and fair. The people of Mali, despite years of authoritarian government, were able to hold multiparty democratic elections in 1992. After one reelection, President Alpha Oumar Konaré stepped down, as was constitutionally required of him, and a peaceful transition of government took place after the elections.

Indonesia, Senegal, and Mali are three postcolonial Muslim countries that stand out for their democratic success. Even though they were not inculcated with the norms and institutions of democracy under colonial rule, they managed to master the intricacies of democratic governance in a short period of time. Democracy after independence did not come quickly, but it did come. The same, sadly, cannot be said for most other Muslim countries. After years of colonial rule, many did not have experience with democracy. In many countries, democratic movements and/or groups sprouted up, only to be crushed by the colonial power or its proxy.

Central Asia

The five Muslim Central Asian republics of the former Soviet Union must be included in any review of the status of democracy in the contemporary world. We should be sensitive to the fact that there are considerable ideological and religious variations amongst these five newly independent states. And when one travels through this area of Central Asia, one is struck by the almost total physical absence of symbols of Islam.

Most of the land that makes up modern-day Kazakhstan was part of the Russian Empire and later, the USSR. In 1991, along with all the other Soviet Socialist Republics, Kazakhstan became independent. Nursultan Nazarbayev, who was the leader of Kazakhstan during Soviet rule, was elected as president in a landslide victory in 1991 in a contest in which there were no other candidates. Nazarbayev has won every subsequent presidential election. In 2005 the Parliament voted to allow Nazarbayev to run for the post of president as many times as he wanted, reversing a constitutional two-term limit. Kazakhstan is a multiparty state, but the president's party, Otan, dominates politics. Multiparty elections are permitted, but opposition parties claim that there are legal obstacles to getting a real chance at power. The most recent elections, in 2007, fell short of international standards, and the Otan party won 88 percent of the seats. Kazakhstan is the richest and most pluralistic of the Central Asian republics, and its market-based economy is the strongest. Despite the constraints of the Nazarbayev era, there are functioning NGOs and growing political parties. Over the next period, Kazakhstan may move dramatically to open up its political process to greater participation.

The Kyrgyz Republic became an independent state in 1991. Askar Akayev, the leader of the state since 1990, was elected president in an unopposed 1991 victory. Throughout the 1990s, the strong president struggled with the increasingly independent Parliament. These struggles became more acute in 2002, when state officials arrested a leading opposition figure and five people were killed by police in the ensuing protests. The turning point for Kyrgyzstan came during the 2005 parliamentary elections. Irregularities and vote rigging prompted widespread discontent, which turned into protests and demonstrations. The government was sacked, and Akayev was forced to resign. This "Tulip Revolution" led to the July 2005 election of Kurmanbek Bakiyev as president. In December 2006 constitutional changes maintained a strong president but codified some of the positive changes of the Tulip Revolution, such as a strong multiparty system, into law.

Tajikistan devolved into civil war almost immediately following independence from the Soviet Union. Although Imomali Rakhmonov was elected to the post of president in 1994, the violence in the state did not end until a cease-fire was signed in 1997. He was reelected in 1999 in elections that a Human Rights report called a "farce." The 2006 presidential elections were a slight improvement but still lacked fairness, according to international sources, including the National Democratic Institute. Governance in Tajikistan mirrors the election process, as the government has all the trappings of a democracy (multiple parties, a constitution, shared power), but in reality Rakhmonov rules in an authoritarian manner with a small circle of advisors.

From Soviet independence until 2006, Turkmenistan was ruled by Saparmurat Niyazov. He was elected in a 1992 election that featured no other candidates. His rule was extended for eight years in 1994, when he received an impossible 99.9 percent of the vote in a plebiscite. In 1999 he was relieved of the burden of having to run in elections, as Parliament declared him president for life. Following his death in December 2006, Gurbanguly Berdymukhammedov, a son of the former president, was elected in a contest that was condemned as corrupt by the international community. He faced only four other candidates, who were from the same (and only legal) party. The president has most governmental power securely in his hands, and press freedom is all but nonexistent.

The former first secretary of the Communist Party in Uzbekistan, Islam Karimov, was elected president in 1991 following Uzbek independence. He was reelected in 2000 and will face reelection in December 2007. Both presidential elections thus far have fallen well short of international standards. Human Rights Watch reported that in the 2000 election no true opposition parties were allowed to field candidates and press freedoms were seriously curtailed. The same complaints were registered in the 2004 parliamentary elections. Much like other Central Asian states, Uzbekistan has all the trappings of democracy, but President Karimov rules in an authoritarian manner.

Middle East

Continuing our democratic *tour de raison,* we turn to the Middle East, where a few states register as "partially free" on the Freedom House scale. Some of these states have rulers in place whose rule was "legitimized" and strengthened by a colonial power. Thus, upon independence, their stranglehold on power was too strong for democratic forces to unseat. Kuwait and Oman fall into this category. Outside the Middle East, Brunei would also be an example of this phenomenon.

Africa

In Africa there has been a pattern of postindependence democracy, followed by a quick coup destroying the new democracy. Usually these coups were carried out by the military. The militaries of many of these states were relatively strong, usually as a result of training provided by a colonial power, so that the army could be used as an instrument of oppression during colonial rule. Gambia, Guinea, Mauritania, and Niger are examples of this.

During the colonial era, imperial powers divided up and consolidated states at will. This lack of respect for ethnic and tribal lines often put two warring peoples into one state or split up a single people into two states. When independence was gained by these states, democracy broke down along ethnic lines and many times proved impossible to sustain. Good examples of this are Lebanon, Somalia, Sudan, and Western Sahara. Iraq was also such an artificial national creation.

Of the remaining Muslim states, Syria and Saudi Arabia have strong central governments without any functioning institutions of democracy. Djibouti, Comoros, the Maldives, Yemen, Sierra Leone, Morocco, and Malaysia have exhibited a history of gradually loosening autocracy and seem to be moving toward progress in democratic development.

The Persian Gulf States

The Persian Gulf is an area of interest because of its increasing Western presence (primarily through the U.S. military) and thus exposure to Western political values. Kuwait has free elections for members of its unicameral Parliament. There are fifty members of this Parliament, and fifteen automatically come from the cabinet—which is appointed by the prime minister, who is appointed by the emir. The Parliament does have substantial legislative powers, as exemplified in its 1999 rejection of the emir's proposal that women get the right to vote. This rejection highlights one of the common trends in the Gulf—a ruling class that is more liberal than the ruled. The ruling Sabah family remains firmly entrenched in power.

Bahrain is another example of suppression of democratic growth in the Persian Gulf. Bahrain was a British protectorate. The British instituted some reforms in Bahrain, such as developing education for girls and building the pearl industry. However, when leftist labor parties developed and demanded political reform in the 1950s, their leaders were arrested and deported. When Bahrain achieved independence, the new Constitution established an elected legislature. However, when differences arose between the new legislature and the emir, the emir dissolved the body. This legislature was brought back by the ruling family in unilaterally imposed constitutional changes in 2002. The changes established a bicameral legislature with an elected lower body and an appointed upper house. The two houses are the same size, which effectively diminishes the power that the lower house had in the previous Constitution. The elections that followed were considered fairly transparent and provided universal suffrage. While the 2005 changes have brought an increased measure of democracy to Bahrain, King Hamad bin Isa al-Khalifa and the ruling family still dominate the country.

Qatar is another Gulf country that is struggling to build a democracy. Qatar became a British protectorate in 1916. Under British control, the government of Qatar was a monarchy. In 1971, when Qatar finally became an independent, sovereign state, it remained a monarchy. There has been some sociopolitical liberalization under the current emir, Hamad bin Khalifa al-Thani. In 1999 the emir allowed local elections in which women could vote and run for office. In 2005 a new Constitution allowed for a unicameral legislature, with two-thirds of the members being elected. This body can even vote government ministers out of office and approve budgets. Elections for this body have been scheduled but not yet held. De facto, Qatar remains a monarchy and Qataris possess few constitutional protections and limited opportunity to participate in the decisions of their government.

President Ali Abdullah Saleh has ruled United Yemen since 1990. He first came to power in 1978 in North Yemen through a coup. Since unification he has ruled as president of a united Yemen. In 1999 the first set of presidential elections took place (Saleh had been ruling as provisional president since unification). He was elected and was re-elected most recently in 2006. He is not eligible for a third term, but when the 2013 elections take place, his son Ahmad will be old enough and is expected to run. Many parties are allowed to contest elections, but obstacles are put in their way. In the 1999 presidential election, the candidate from the main opposition party was disqualified from running. Although reforms have taken place under Saleh, most of his efforts go toward keeping consolidation of his power.

Saudi Arabia is a strict monarchy, where absolute power still lies with the king. In recent years, however, there have been slight moves toward the beginnings of democracy. In 1992, in response to citizens' request, King Fahd passed the Basic Law, which outlined citizens' rights and established the Shura Council. This assembly is purely consultative, and all its members are appointed by the king. The members are, however, all nonroyal, and political discussion is held by the Shura Council. The body even includes two Shia Muslims, traditionally a repressed class in the kingdom. In 2005 King Abdullah went a step further and allowed half of the municipal counselors to be elected. The electorate was limited to males and Muslims. Discrimination against women is often statutory.

.

Western intervention for perceived self-interest that undermines emerging developing world democracies has not, of course, been limited exclusively to the Muslim world. There is a long colonial record of resource exploitation and depletion, and an equally long and sordid pattern of interventions to block nationalist movements that threatened Western economic or political goals. Let's briefly review four classic case studies.

Guatemala

An example of Western decapitation of an emerging democracy that led to a long period of autocracy is the Central American country of Guatemala. In the 1930s Guatemala was ruled by a repressive leader named Jorge Ubico Castañeda. Following a 1944 revolution, Guatemala held fairly contested democratic elections and a progressive candidate named Juan José Arévalo was elected president. He enacted a series of land reforms and labor laws that threatened the interests of the famous American-owned United Fruit Company. United Fruit was the largest company in Guatemala, employing 40,000 workers. It bought controlling interests in railroad, electric, and telegraph industries, and took control of the nation's only port. It was notorious for treating workers extremely poorly, with a pattern of withholding taxes that were never transferred to the government but kept as profits.

The Arévalo regime passed significant changes to the nation's labor code, giving workers the right to organize and petition for grievances in salary and working conditions. Workers almost immediately organized into a powerful union against United Fruit and went on a strike that shut down the company. The leadership of the company appealed to the U.S. State Department for help. United Fruit also began a rather sophisticated public relations campaign to undermine President Arévalo and began funding Carlos Castillo Armas, a rightist, with weapons and money.

Initially President Harry Truman resisted the calls for U.S. intervention in support of United Fruit, and supported the democratically elected President Arévalo. But after the fall of China to communism and after the Soviet Union tested nuclear weapons for the first time, the CIA and the State Department determined that President Arévalo's government had been infiltrated by Communists and that these Communists threatened to take over the entire country. It is difficult to substantiate the accuracy of the assessment or to portray it as a paranoid reaction to the spread of communism in other parts of the world. In any case, American policy toward Guatemalan politics became much more aggressive.

In 1950 Arévalo was defeated by Jacobo Arbenz Guzmán in a democratic election that marked the first peaceful democratic transition in Guatemalan history. The experiment with democratic stability and continuity was short-lived.

After President Arbenz began confiscating land owned by the United Fruit Company, the attitude of the United States toward his regime changed dramatically. The United States became convinced that the Arbenz regime was directly tied to the Communist Party of Guatemala, the PGT, and that Guatemala could very well become a Soviet beachhead in the Western Hemisphere. Three years after the democratic election of President Arbenz, the U.S. Central Intelligence Agency, collaborating with the Nicaraguan dictator Anastasio Somoza Debayle and the Dominican dictator Rafael Trujillo Molina, designed and executed a covert operation that took down the Arbenz government and replaced it with a regime under the rightist Carlos Castillo Armas.

Armas instituted a brutal and repressive regime and canceled the presidential elections that were scheduled for 1955. Yet U.S. vice president Richard Nixon visited Guatemala and praised its "democratic government." This public accreditation of an unabashed and brutal dictator as a "democrat" had a long and bitter impact on Guatemalans' attitudes toward the United States. In 1958 the unpopular Armas was assassinated. The military's candidate, who was supported by the United States, lost the subsequent presidential election, but the army seized control of the government and anointed the losing right-wing presidential candidate, Miguel Ydigoras Fuentes, president of Guatemala. The United States immediately recognized his legitimacy.

Greece

During World War II, Greece became a close ally with Great Britain, which supported the Greek government in exile in London. As the war came to a close, it became clear that the government in exile of King George II would not be able to easily assume the reins of power. In Greece, the umbrella leftist party, EAM, was demanding a large role in the future Greek government, as its military groups had played prominent roles in Greek resistance to Axis rule.

As the Nazis withdrew from Greece in 1944, the leftist parties refused to disarm. British prime minister Winston Churchill then sent British troops to support the Greek government under George Papandreou. This brought a tenuous peace to the country, until full-blown civil war broke out in 1946. Under an agreement reached earlier between Churchill and Stalin (without Greek input), Greece was to be placed under mostly British influence. Thus, Great Britain, and later the United States, felt justified in trying to keep Greece in the non-Communist camp. Showing the importance the United States put on keeping Greece non-Communist, President Truman sent $200 million in aid to Greece in 1947 to aid the government in its fight against the Communists. The civil war was finally won by the rightists when Josip Broz Tito, the ruler of Yugoslavia, closed the border between Yugoslavia and Greece, depriving leftist forces of the support they needed.

After the Civil War of 1946–1949, conservative governments ruled Greece. During that time the United States assisted the Greek security and intelligence apparatus in destroying the Communist Party of Greece, which had been outlawed after the Civil War. Everything came to a head when in 1967 the military overthrew the civilian government in response to a possible left-center coalition government. The ruling military junta proclaimed martial law. Political freedoms and civil rights were dissolved or severely curtailed.

The legacy of U.S. (and earlier British) influence in Greece was years of political instability capped by a repressive dictatorship. U.S. officials felt that Greece was a vital state in the fight against communism. Thus, starting with the Truman Doctrine in 1947, American policies toward Greece were centered on an anti-Communist agenda and did little, if anything, to promote democratic values there. Successive U.S. governments favored rightist dictatorships and the repressive Greek security apparatus over promoting democratic governance.

Argentina

Argentina was a struggling democracy throughout the twentieth century. By any measure, it had a number of fair, democratic elections. Unfortunately, these elections were usually separated by years of military rule. A turning point came in 1973, when the popular political figure Juan Perón returned from exile. His return was marred with violence as the Montonero party, a Peronist right-wing group, and the Alianza Anticomunista Argentina party, a Peronist left-wing group, killed more than a dozen people and wounded hundreds at Perón's airport reception.

The ensuing resignation of then-president Héctor José Cámpora allowed Perón to run in and win democratic elections of 1973 with 61 percent of the vote. Tragically, however, he soon died of a heart attack and his vice president and second wife, Maria Estela Isabel Martínez de Perón, was unable to keep the country together amid growing terrorist violence and economic instability.

In 1976 a military coup toppled her government and began a "process of national reorganization." This process involved political repression on a national scale. Peronistas were purged from the government, and members of left-wing terrorist groups were targeted by the government. In the government's quest to destroy these groups, thousands of innocents (as many as 19,000) were killed in government-sponsored killings, which included unlawful kidnapping, detention, and torture.

In 1979 the U.S. government under President Jimmy Carter, who had made human rights a centerpiece of bilateral relations in his administration, suspended loan negotiations with Argentina in the face of massive human rights abuses by the Argentinean government. When President Ronald Reagan came into power, the situation was reversed. During Reagan's presidency, the CIA was not only aware of the abuses by the military government but worked with Argentine intelligence officers and had Argentine intelligence officers train the Nicaraguan contras. This is yet another unfortunate example of the United States supporting a nondemocratic government and even supporting political repression on a large scale to further a political agenda.

Democratic Republic of the Congo

The Democratic Republic of the Congo was the personal property of the Belgian monarch, Leopold II, beginning in 1885. In 1908 the Free Congo State, as it was then named, became a colony of Belgium. After World War II, tensions began to rise in Belgian Congo as many colonies around the world and in Africa began to take steps toward achieving independence.

In the late 1950s two Congolese groups—the MNC, led by Patrice Lumumba, and ABAKO, led by Joseph Kasavubu—took center stage in the Congolese fight for independence. When the Belgian government announced in January 1959 that at the end of the year there would be local and national elections, the race was on for power in the Congo. The June 1960 elections gave Lumumba a plurality, and he formed a coalition government with Kasavubu as his president.

Unfortunately for the young state, unrest broke out immediately. Belgian troops were airlifted in to keep the peace but soon ended up protecting the mineral-rich, seceding territory of Katanga led by Moise Tshombe. Prime Minister Lumumba issued an ultimatum to the United Nations: he needed help with the ejection of Belgian forces, or he would turn to the USSR for help. This action did succeed in getting the United Nations to send troops but caught the attention of the United States, which became convinced that Lumumba was threatening to turn Congo into a Communist state.

Soon the Kasavubu-Lumumba coalition began to fail and Kasavubu removed the prime minister from office. In a fast-paced series of events, General Joseph-Désiré Mobutu, after a failed first coup, took over in November 1965, renaming the country Zaire. During the first coup attempt, Lumumba was arrested and murdered with Belgian help and the CIA's acquiescence. The United States was one of the first countries to recognize the new regime.

Mobutu, supported by the West, soon consolidated power into his own hands. He removed legislative powers from the parliament and assumed full powers. Soon he outlawed all political parties but his own, eliminated local assemblies, and repressed all opposition.

In 1990, under pressure, Mobutu agreed to a multiparty state, but after a 1992 power-sharing agreement, elections were never held. It wasn't until the violence in Rwanda spilled over into Zaire that Mobutu was finally ousted in the First Congo War by Laurent-Désiré Kabila in 1997. The country then went back to being called the Democratic Republic of the Congo.

Soon, however, many of Kabila's former backers turned against him, and the Second Congo War started, which went on until 2003. The wars were some of the bloodiest since World War II, killing almost four million people. Today Congo stands on the brink of democracy, having recently seen violence during its national elections The case of Congo is similar to that of other postcolonial states. A Western power (in this case Belgium) claimed the area as a colony and did not allow any sort of democratic values to develop. When independence was finally reached, Belgium looked out only for its interests and, through military intervention, contributed to the downfall of a nascent democracy. The United States played a role as well, supporting a coup and then the subsequent dictatorial regime that followed instead of the democratic government that it replaced. The West's interests, as always, were focused on its own mineral needs and its anti-Communist foreign policy stance, to the detriment of the Congolese people.

·

I have talked about a century and a half of history of experimentation and practice of democracy within the Islamic world. It has been a rather complex and imposing task to look for patterns of behavior that might help to explain the development—or more frequently the lack of development—of democratic institutions, processes, and governance in nations with majority-Muslim populations. It is certainly a mixed record of starts and stops, of progress forward and regression backward, of missed opportunities and roads not taken. There is nothing in this historical rendition that supports the notion that Islamic values and the democratic experience are inherently mutually exclusive. Under the right conditions, with the right support and the nurturing of time, we have seen the development of significant democratic infrastructure and institutions within majority-Muslim nations. But those right conditions have been all too infrequent, and support from outside has been often very little and very late, often too late to protect young and developing democracies from internal destabilization by regressive forces within their own societies.

It was for this very reason that I proposed, in my commencement speech to the Harvard graduating class of 1989, the creation of an Association of Democratic States. This concept, which has developed into the Community of Democratic Nations, seeks to sustain and empower vulnerable new democracies from internal and external threats to their evolving democratic infrastructure. It needs to be strengthened.

Few could argue that the colonial experience has not had a deadening effect on the development of democratic practices, elections, and governance throughout the Muslim world. Colonial powers such as Great Britain, France, Portugal, and Holland rarely practiced what they preached in the colonies that they have governed. By definition, colonialism and imperialism are antithetical to the values of democracy, liberty, and freedom. Colonialism is predicated on subjugation. Countries that thrived on democratic constitutions, competitive political parties, vibrantly contested elections, an open and free media, an independent judiciary, autonomous and important NGOs, adherence to parliamentary norms and processes, separation of powers, and constitutional protection of minority rights and human rights clearly did not apply those democratic values to the colonies they occupied and exploited.

On the contrary, although natural and human resources were drained from the colonies, there was little effort to inculcate the values of democracy that could have built an infrastructure for democratic independence. More than merely failing to inculcate and proselytize democratic values, it is clear that colonial powers feared that the spread of democratic norms would lead rather directly to nationalist movements and demands for independence. Nations that were economically dependent on the resources of colonies believed they could not afford the risk and consequences of encouraging democracy in their colonies.

This does not explain or obviate the responsibility within societies with majority-Islamic populations to assert democratic aspirations and apply democratic institutions to the values within their own societies. A child who is not properly cared for and is raised by abusive parents must, at some point in life, assume responsibility as an adult for his or her own actions and future. It serves no one's interests to put all the blame and responsibility for the lack of democratic development in the Muslim world on the backs of colonialism or superpower politics during the Cold War. Was it part of the problem? Certainly. Was it exclusively the cause of democratic atrophy within Islamic societies? Definitely not. For every Operation Ajax, there were a shah and a military that became the instruments of political oppression and human rights abuse. The CIA may have needed the cooperation of the ISI in Pakistan to conduct a surrogate war against the Soviet Union in Afghanistan. But it was General Zia-ul-Haq and his extremist allies within the military, intelligence, and religious communities who perpetrated a reign of terror against the people of my homeland.

And even if the West bears significant responsibility for the lack of democratic political development within the Islamic world, at some point responsibility and accountability rest with us. If democracy is to take hold among the billion Muslims on this planet, the movement must come from our own people standing up to the forces of extremism, fanaticism, and authoritarianism within our own societies.

Of course the historical interface of Western powers and emerging Islamic democracies is not just a function of the economic exploitation of the colonial era. The problem with the West is one not just of omission but also of commission. For a century the issue has not been simply benign neglect but something even more disturbing and potentially even more dangerous. More problematic than colonial powers not applying democratic values to the political institutions of the colonies is a specific and unfortunate pattern of intervention by Western powers in country after country to disrupt and undermine the growth of nascent democracies. This pattern has obviously not been limited to developing Muslim democracies, but we have focused on them in this investigation of Islam, democracy, and the West. The litany of interventions to disrupt the growth of democracy is demoralizing, especially in light of today's political environment.

Actions have consequences, and the actions that the West has taken over the last century to discourage the development of democratic institutions in the Muslim world are clearly related to the cynicism of Muslims as to the true motives of the West and the charge of hypocrisy leveled against the West that is prevalent in the Islamic world today.

How would the world be different if Britain and the United States had not destabilized the democratically elected government of Mossadegh in Iran in 1953?

How might events be different today if Britain had worked toward building a democratic Iraq, allowing political parties to flourish and elections to move into democratic governance, protecting minority rights while preserving majority rule?

And how would the world be different today if the West, and in particular the United States, had not used Afghanistan as merely a blunt instrument to trigger the implosion of the Soviet Union? How would Afghanistan, Central Asia, the Middle East—indeed, the entire world—be different if the CIA had not colluded with the Pakistani ISI to train and empower the most extreme, fanatic, intolerant, and Western-hating elements of the mujahideen coalition, and instead had worked to empower those elements that not only opposed the Soviet occupation but valued pluralism, consultation, and accountability in government—all basic tenets of the norms of prescribed governance in Islam.

And how would the world today be different if, after the Soviet defeat and withdrawal from Afghanistan in 1989, the West had not abandoned that war-torn and beleaguered nation but rather had chosen to help rebuild Afghanistan both physically and civilly, building infrastructure, investing in education, health, and housing, giving farmers incentives to grow crops other than poppies, making a long-term commitment to the Afghan economy, polity, and society?

None of these questions is raised as an excuse for the internal culpability within Islamic societies, for the failure of democracy to take root or as a rationalization for the brutality and unspeakable human rights abuses of Muslim dictatorships. The world would be very different if the West had been more prudent in its actions, had adhered to its own democratic values, and had taken a long view of history rather than a myopic short-term view. If the West had realized that "the enemy of my enemy" is not always my friend, repressive regimes in Iran, Iraq, and Pakistan might have been forced to reform and democratize, knowing that they would not be sustained indefinitely by the superpower support. But again, that does not obviate the hard fact that the actual repression and abuse was internal, Muslim against Muslim. There is enough blame to go around for both the West and the Islamic world. It is shared responsibility and often shared failure.

The political, economic, and social prerequisites for xenophobia that manifest themselves in the growth of religious fanaticism, extremism, and international terrorism may have links—directly or indirectly—to decisions made, actions taken, and choices selected by the great Western powers in dealing with the Islamic world. There has been too little vision. The West has been not proactive but reactive. It has far too often been content to preserve the status quo and not build a pragmatic, sound, and stable future. And in that decision may lie some of the roots of the terrorism and extremism that now threaten the world order and the values of civilized society. Saying that, however, in no way minimizes or excuses Muslims' culpability for internal repression of our own people and the export of international fanaticism and terrorism. The West may have made mistakes, but it is the fanatics and extremists, training a successor generation of terrorists, perverting and distorting the fundamental values of Islam, in safe harbors in the tribal territory of my own nation, who are also responsible for the contemporary assault on the values of the civilized world. And these values are of neither East nor West—they are universal.

After the defeat of totalitarianism in World War II, the United States made an extraordinary commitment to freedom and liberty. It was also a pragmatic calculation about the nature of humanity. A country that had bled profusely and invested human and material capital on the battlefields of Europe and the great seas of the Pacific courageously chose not to retreat from international responsibility.

The United States could have turned inward. Yet it chose another path.

Though only 15 percent of the American people (according to a Gallup survey at that time) wanted to continue the United States' commitment to Europe after 1945, a great president with a great vision recognized that the most effective way to contain the spread of communism in Europe was to rebuild the economies of the nations of the continent. The Marshall Plan was predicated on the concept that people, whose social needs are being adequately addressed by government, will neither acquiesce to nor tolerate totalitarian or extremist hegemony. President Harry S. Truman invested billions in rebuilding Greece, Italy, and France. He committed the money and prestige to not only rebuilding the defeated Germany and Japan economically but nurturing and supporting viable and thriving democracies. This vision and methodology—more so than even the great military of NATO—is what ultimately defeated communism. As Václav Havel, the former president of Czechoslovakia and later the Czech Republic, said so presciently in 1992, "Communism was not defeated by military force, but by the human spirit, by conscience, by the resistance of man to manipulation. It was defeated by a revolt of human individuality against imprisonment within a uniform ideology."

The world would be a very different place if the West had made a similar commitment to building the economic and democratic political infrastructure of the Muslim world instead of frequently draining its material and human resources and thwarting the growth of democratic values. An educated society—where parents have jobs, where children are properly fed, housed, and clothed, where a successor generation has hope to build a better life than their parents'—is not the kind of society, not the kind of political and social environment, that would be susceptible and vulnerable to the appeal of religious fanatics and political extremists, nor would it accept the legitimacy of a regressive and internally interventionist military as a key political power player within society.

Hunger breeds extremism. Hopelessness breeds extremism. Despair breeds extremism. Opportunity makes democracy thrive. A government that addresses the human needs of its people is given the benefit of the doubt. A government that successfully addresses the day-to-day needs and concerns of its people is likely to be given the necessary political support and the time to grow democratic structures. People who are deprived are impatient, and that impatience is often radicalizing. Nations with Islamic majorities may greatly benefit from assistance from the West to make the necessary progress of social and economic development that will sustain progress on political development, although ultimately the responsibility rests within the Islamic world.

These are some of the key lessons that we can take from reviewing the state of democracy in the Muslim world through history, and today. We will apply these lessons later in this treatise as we attempt to shape and mold a realistic framework to reconcile the peoples, values, and institutions of the West and the Islamic world.

4

The Case of Pakistan

A s I write these pages, Pakistan is in great turmoil. In its sixty-year history, November 3, 2007, will be remembered as one of its blackest days. Pakistan is currently a military dictatorship. On that date, General Pervez Musharraf removed all pretense of a transition to democracy by yet another extraconstitutional coup d'état. He suspended the Constitution and arrested hundreds upon hundreds of party officials, human rights activists, lawyers, judges, and journalists. He suspended independent television. He banned print media that are critical of his military rule. He ordered the police and the military of my nation to baton-charge, beat, and tear-gas our own people. It was a black day for Pakistan, a black day for democracy, and a black day for the more than one billion Muslims around the world who look to Pakistan as a model of moderation, civility, progress, and democracy. I now look at how Pakistan reached this point by tracing the history of the country.

To understand Pakistani politics, an understanding of Pakistan's provinces and their characteristics is necessary. Pakistan has four provinces: Punjab, Sindh, Baluchistan, and the North-West Frontier Province. Additionally, it has four separate federally administered areas, including the capital, Islamabad; Pakistan-controlled Azad Kashmir; the Northern Areas; and the Federally Administered Tribal Areas (FATA), which are directly governed by the federal government. FATA is located between the North-West Frontier Province and Afghanistan. Under Musharraf, the authority of the government has collapsed in FATA. Local tribal leaders and religious militias dominate the area and make it a haven for extremists. The Northern Areas are beset with sectarian violence and bloodshed. Azad Kashmir is autonomous. Its final status will depend on a U.N. plebiscite. Pakistan claims both territories as well as the Indian-held Jammu and Kashmir.

The Punjab is Pakistan's largest province in population, where half the nation's people (88 million) live. Punjabis are the main ethnic group, but there are others. Punjabi, Urdu, and Seraiki are spoken in Punjab. Punjab means "five" (*punj*) rivers and, as the land of the five rivers, is most fertile. The Punjab is well developed agriculturally and industrially, growing such crops as wheat, gram, bajra, raic, corn, cotton, and sugar cane. Its industries include textiles, mining, and some cottage industries such as metalworking and carpet weaving.

Sindhis form the main ethnic group in Sindh, where the Sindhi language is spoken. The Sindh gave refuge to large numbers fleeing India when the nation was born in 1947. They are known as Muhajirs. Karachi, the provincial capital, is Pakistan's most important industrial city and commercial capital; it is also Pakistan's most important port. Named after the mighty Indus river, Sindh has a thriving agricultural sector. It has developing industries in finance, textiles, food, chemicals, metalworking, and information technology.

Baluchistan, Pakistan's largest province by land area, has just 8 million people living in the mostly arid plateau. The Baluchis can boast of one of the biggest natural-gas fields in the world. From time to time Baluchistan has been racked with violence associated with a movement for independence. During my first government, the Pakistan Peoples Party began building a series of ports on the Mekran coast, including Gwadar. My father had visualized these ports as new trading routes with Central Asia.

The religious parties have gained in strength in Baluchistan following the influx of Pakhtun refugees from Afghanistan following the Soviet invasion of 1979. Baluchistan shares a border with Afghanistan. If development is a priority, Baluchistan is a key target for PPP resurgence in any free and fair election in the future.

The North-West Frontier Province is the smallest province by land size. It is here and in FATA that the battle against religious extremism is most concentrated. The majority of the people are, like those across the border in Afghanistan, Pakhtuns. Although the MMA religious coalition formed the provincial government in the rigged elections of 2002, the Leader of the Opposition belongs to the PPP, which has a significant following in the province.

Pakistan's modern history traces itself to the Muslim rulers of undivided India, the greatest of whom were known as the Mughals. The Mughals united the largest number of small princely states and kingdoms during their rule. Babur, the founder of the Mughal Empire, was a descendant of Genghis Khan. He came to South Asia from Farghana in Central Asia through Kabul in the early sixteenth century. After capturing most of modern Afghanistan, he went into the Punjab and from there to Delhi, which he conquered in 1526. His son Humayun took over in 1530. He spent his life in conflict, fighting to keep his kingdom. By 1555, near the time of his death, he had reconquered Delhi. His son Akbar inherited the Mughal Empire in 1556. He was the greatest of the Mughals, who expanded and modernized the Mughal Empire. He promoted religious tolerance in the vast land that included people from the major religions of Islam, Buddhism, and Hinduism, as well as a myriad of local religions. He ruled this vast sea of humanity not by forced conversion but by the power of pluralism and multiculturalism.

The Mughals were a Muslim minority in largely Hindu India. Yet Akbar united his kingdom through an inclusive set of values, reaching out to embrace all subjects as equal in the eyes of government and law. He married women from leading Hindu families in his successful bid to unite large tracts of the religiously and ethnically divided India.

During Akbar's reign and throughout the empire's lifetime, the Mughal Empire contributed much to the modern world. The Mughals believed in education for both men and women. The princesses of the court were well versed in poetry, politics, painting, and history. Major accounts of the life and government of the Mughals survive through the autobiographies of the kings, princesses, and courtiers.

Mughal women were strong personalities who had a say in government through their husbands and fathers. The beautiful empress Noor Jehan was said to have ruled the Mughal Empire during the reign of the emperor Jehangir. Her father had fled Iran, or Persia, as it was then known, following the death of a king and the war of succession that followed it. Her father escaped from Iran to New Delhi. On the way at Qandahar, in Afghanistan, was born Noor Jehan, who would rule India with an iron fist. Ironically, many centuries later, the headquarters of the Afghan Taliban warlord, the one-eyed Mullah Omar, would be Qandahar, the city of the birth of one of Islam's most powerful women leaders.

The Mughal ruler Shah Jehan (grandson of Akbar) built one of the Seven Wonders of the World, the Taj Mahal. It is a shimmering white marble mausoleum built in memory of his wife, Mumtaz Mahal, the niece of Empress Noor Jehan. Many cite it as a lasting monument by a loving husband toward a loving wife. The cost of building the Taj Mahal and other architectural monuments and gardens crippled the

Mughal economy. The resplendent grandeur of the reign of Emperor Shah Jehan, who also built the Shalimar garden in Lahore, gave way to the austere rule of his son the Emperor Aurangzeb.

The last great Mughal emperor, before the empire went into decline, was the religiously driven Emperor Aurangzeb Alamgir (1618–1707). In his zeal to spread Muslim might to the farthest corners of Hindu India, he overextended his rule. By so doing he planted the seeds of its decline. His successful campaign against the Marathas was also the beginning of the decline of Mughal power. While victorious, this costly war opened his empire to attacks from other groups including the Afghans and Persians.

Coinciding with the birth of the emperor Aurangzeb, British forces reached India in the early seventeenth century. The British East India Company began signing treaties with local leaders and the Mughal emperors. In 1617 the East India Company was given trading rights by Emperor Jahangir.

As the Mughal Empire overextended, its economy weakened by constant war and extravagance, the East India Company, a rich trading group, moved in to fill the power vacuum. Guns and money speak, and so the British began annexing territories, promising local rulers both protection and expanded trade. As trade expanded, so did British influence. At its zenith, the British ruled across the Subcontinent, including the modern nations of Pakistan, India, and Bangladesh. By the nineteenth century, the Mughal rulers were reduced to figureheads. In 1857, a revolt took place against British rule and many British were massacred. The British narrowly regained control and deposed the last Mughal emperor, Bahadur Shah Zafar. He died in exile, in poverty and isolation, lamenting that he, a king who had ruled a vast empire, did not now own even a few feet of it to be buried in. With the overthrow of the last Mughal emperor, India ceased to be part of the East India Company. It became part of the British Empire on which the sun never set, for it stretched across continents and time zones.

In 1843 the British conquered Sindh, including the Indian Ocean port of Karachi, vital for trade routes. Then they marched on and captured Baluchistan. Twenty-eight years after the Indian Mutiny, or the War of Independence, as South Asians call it, the Indian National Congress was formed in 1885. It was the first political party created on the Subcontinent. Through peaceful means the Congress fought for local self-government. In 1906, the Muslims of the Subcontinent formed the All India Muslim League. The All India Muslim League aimed to protect the religious, cultural, economic, and political rights of the massive Muslim population on the Subcontinent. Mohammad Ali Jinnah, the founder of Pakistan, initially joined the Indian National Congress. When the Muslim League was formed, Mr. Jinnah was a member of both the Muslim League and the Congress. Both organizations permitted overlapping membership. However, in 1920, Mr. Jinnah left the Congress. Soon he became an important leader of the Muslim League.

Interestingly, the party cobbled together by General Pervez Musharraf through Pakistan's secret services took the name Muslim League as well. However, its ideology is different from that of Jinnah's Muslim League. Using the same name was an attempt to connect with the country's founding political movement.

In 1916 the Indian National Congress and the Muslim League signed the Lucknow Pact. Under this pact the League would support the Congress in its effort to get independence for India from the British. In return, Muslims would have special seats in the Indian Parliament known as separate electorates. Muslims wanted special seats separately allocated in the hope of having stronger representation in Parliament than that obtainable through direct elections, in which Hindus often had the majority due to their population size.

The Muslim movement for a separate identity took a dramatic turn in 1930 in the city of Allahabad. Pakistan's poet-philosopher, the secular leader Mohammad Iqbal, addressing the Muslim League, called for the creation of an autonomous Muslim state.

My grandfather, Sir Shah Nawaz Bhutto, had formed the Sindh United Party in the 1920s. He had also launched Sindh's first newspaper, *The Zamindar.* During British rule only those who paid taxes were allowed to vote for local governments. These were often lawyers and landlords. All others were disenfranchised.

Throughout history, Sindh had been an independent empire. Once its capital had been Multan, a city now in Punjab. At its height, Sindh extended all the way north to Afghanistan until the British amalgamated Sindh into India. There were three main "states": Sindh, Hind (India), and Bengal, built on the banks of the Rivers Indus (also known as Sindhu), Ganges, and Brahmaputra, respectively.

The British conquered Sindh in a ferocious battle at Miani in 1843. After conquering Sindh, the British general, Sir Charles Napier, telegraphed the Latin word *"Peccavi"* to London. Translated, it meant "I have sinned," a play on the word *"Sindh."*

The British ended Muslim Sindh's separate identity by integrating it into Hindu-majority Bombay. My grandfather, as representative (or mayor) of the Larkana Board, had long struggled for the separation of Sindh from Bombay. The British said that the waterlogged, saline Sindh lacked sufficient revenues to be independently governed as a separate administration.

My grandfather then initiated the Sukkur Barrage project to turn the arid lands of upper Sindh fertile. With the completion of the Sukkur Barrage, Sindh gained sufficient revenues for my grandfather to argue that Muslim Sindh be separated from Hindu India. He was successful, and Sindh once again emerged as a separate entity under British rule.

My grandfather and Mr. Jinnah both studied at the Sindh Madrassa, the first modern school established in Sindh's capital, Karachi. The Sindh Madrassa was built by Haji Effendi, the maternal greatgrandfather of my husband, Asif Ali Zardari. Mr. Jinnah and my grandfather lived near each other on Malabar Hills in Bombay. Mr. Jinnah stayed with my grandfather at our house in Larkana on his visits there.

By separating Muslim Sindh from Hindu India, my grandfather played a critical part in what was to become the Pakistan Movement.

In 1935, Sindh elected its own Assembly when the British, conceding partial self-rule, allowed elections to provincial parliaments while the central government remained firmly in its control. Five years later, in 1940, the Muslim League called for an independent Pakistan composed of the Muslim majority areas of undivided India. The call for the independent state of Pakistan was made in the city of Lahore, the capital of the provincial parliament in Punjab. It is known as the Lahore Resolution. The Muslim League and the Indian National Congress both wanted independence from British rule. However, the policies the two parties charted out were two different courses for the future of the South Asian Subcontinent.

Seven years later, in 1947, Lord Louis Mountbatten, the last viceroy of British India, announced the Mountbatten Plan, which would largely divide the Indian Subcontinent into Hindu India and Muslim Pakistan. Announced on June 3, 1947, the Mountbatten Plan was to give independence in June 1948. The Indian Independence Act was passed by the British House of Commons, on July 18, 1947. Once the British decision to leave India was made, Mountbatten thought it futile to stay on and moved the date of independence to August 1947. Hardly any time was left to prepare for the founding of the two nations. A man who had never been to the Indian Subcontinent, Cyril Radcliffe, was appointed by Mountbatten to draw the borders between the two nations in just over a month. The hasty division led to the demographic chaos that followed.

With the political parties accepting the Mountbatten Plan for partition and the transfer of power, August 14, 1947, was set as Independence Day. Astoundingly, this was only a month after the Indian Independence Act had been passed by the British Parliament. The expedited British departure left the two newly created states with almost no time to organize a peaceful partition or transition-of-power plan for their new governments.

The partition plan gave all provinces and princely states the choice as to which new country, Pakistan or India, they would join, or "accede to." In most territories that did not present any problems:

Hindu-majority states acceded to India, while Muslim-majority states acceded to Pakistan. But a huge problem emerged concerning the status of the princely states of Junagadh, Hyderabad, Kashmir, and Jammu. The Muslim princely rulers of Junagadh and Hyderabad acceded to Pakistan even though they had Hindu-majority populations. The Indian Army then moved in to forcibly annex Junagadh and Hyderabad.

The Hindu ruler of Kashmir acceded to India despite having a Muslim-majority population. The Muslim population revolted. Kashmir ended up divided, with parts controlled by both India and Pakistan.

The hastily arranged partition of the Subcontinent resulted in chaos, violence, and confusion, especially in the areas contiguous to the newly drawn borders. An estimated 20 million people crossed the border from one side to the other—Hindus to India and Muslims to Pakistan. The newly created nations were unable to cope with the staggering flood of refugees, and the violence grew even more intense. In just the first weeks after Partition it is estimated that more than 200,000 people were killed.

After the initial fighting in Kashmir, the issue of the status of Jammu and Kashmir was presented to the U.N. General Assembly in the first days of 1948. India accused Pakistan of trying to seize Kashmir illegally. Pakistan said that the people of Kashmir should decide their future as India had seized the states of Junagadh and Hyderabad on the grounds of their Hindu-majority population, whereas the population of Kashmir was of Muslim majority. While the issue was still being decided by the United Nations, India launched a full-scale war against Pakistan. Although Pakistani troops were able to secure Azad (free) Kashmir, Indians occupied the rest of Jammu and Kashmir. Both sides accepted an August 1948 U.N.-brokered cease-fire and a resolution calling for a plebiscite. But after the war Indian prime minister Jawaharlal Nehru backed out of the U.N. plebiscite, claiming the time was not right to hold a plebiscite in the region.

The first war between India and Pakistan in 1947–1948 high-

lighted some of the handicaps with which the Pakistani state started
off at its birth. At independence, Pakistan and India were to divide
their resources equally, but this in fact never occurred. As an example,
of the 133 civil servants in the Bengal region, only one came to work
in Pakistan. By early 1951, fully 10 percent of the population of Pakistan was Muslim refugees from India, largely from the Muslim minority areas of India. When the military was split, Pakistan received
only six of the forty armored regiments of the army, only eight of
forty artillery units, and only eight of twenty-one infantry units. At
the time of Partition, Pakistan, which was supposed to have received
39 percent of the Indian British Army, received only 30 percent of the
army, 40 percent of the navy, and 20 percent of the air force, thus putting it at a distinct military disadvantage with the Indians.

Compounding these military and bureaucratic inequities were
natural resource issues between Pakistan and India, most notably the
Indus Valley water dispute. The British had built a major water irri-
gation system in West Punjab that had turned the desert area into
fertile harvest land. At Partition, most of the control points for the
canals and the five rivers that fed the canals were in Indian territory.
India immediately shut off the water to Lahore and the surrounding
lands and demanded payment from the Pakistanis for the water that
was vital for Pakistan to function economically, socially, and politically. A quick series of negotiations began to resolve the issue and restore the flow of water, but there was not a final resolution to this
dispute until 1960, a dozen years after the initial crisis.

Partition also left Pakistan economically vulnerable. The only major industry in East Pakistan was jute farming. There was not a single
jute-processing mill in East Pakistan. Under British rule, all the pro-
cessing mills had been built in Calcutta. With no economic relations
between the two hostile countries, the Pakistani jute-farming industry was unable to get its product processed. And in West Pakistan,
there was a similar crisis in cotton farming. Of the 394 cotton mills
in the Subcontinent, only 14 were located in Pakistan. At the time of
Partition only 1 of 57 of the Subcontinent's top companies was owned

by a Muslim. And although Pakistan had a quarter of the land of the Subcontinent, it had only a tenth of the industrial base. Pakistan received only 17 percent of the revenue of pre-Partition India. India was in a vastly advantageous position to compete internationally and to build upon its young democratic institutions.

Political problems made the transition even more difficult in Pakistan. Most of the Muslim League political leadership came from the Muslim-minority areas of India, not from the Muslim-majority areas of modern-day Pakistan.

On top of this, Pakistan was confronted with geographic bifurcation, as it was made up of two geographically separate parts: East Pakistan (now Bangladesh) and West Pakistan (now Pakistan). These two parts were separated by a thousand miles of hostile Indian territory and made up of ethnically different populations.

Many people point to India and Pakistan today and the economic and democratic differences between the two countries. They argue that since both states were part of the same British colonial experience with the same beginnings and circumstances of independence, Pakistan's democratic failure must lie in the only significant difference between the two countries: religion. I find that thinking simplistic and flawed.

The major setback to Pakistan was twofold. First, the founder of Pakistan died a year after independence, leaving the country without the moral authority of a leader above all others who could give a firm foundation to democratic practices. Second, under provincial self-rule, the Indian National Congress had governed several provinces in India and had a grassroots organization. This was not the case in Pakistan.

Pakistan was also handicapped in other key areas, which contributed to the failure to institutionalize democracy. First, the 1948 war with India made Pakistan feel vulnerable to the Indian threat. Consequently, a large portion of the budget was spent on defense to counter India's military. India's military, of course, was backed by a much larger population and economy. Between 1947 and 1950, approximately 70 percent of the Pakistani budget was spent on defense. As

Liaquat Ali Khan, Pakistan's first prime minister, put it, "The defense of the State is our foremost consideration . . . and has dominated all other governmental activities. We will not grudge any amount on the defense of our country." This began the process of giving the military inordinate stature and influence in Pakistani society, while diverting money away from economic and social development. Its political effect was dramatic. Instead of strengthening democratic institutions and infrastructure, unelected institutions, such as the army and the intelligence agencies, took precedence. They became the central institutions of the new Pakistan.

The second critical difference in the democratic development of these two countries was the longevity of their central political figures. India had the advantage of having its father of independence—Prime Minister Jawaharlal Nehru—live long enough to establish a genuine working democracy. Pakistan's founding father, the Quaid-e-Azam (Great Leader), Mohammad Ali Jinnah, tragically died in September 1948, in the first year of Pakistan's life, long before his work was done, indeed just as it was beginning. Strong leadership supported by a popular mandate is crucial for any state, especially a new state. Pakistan's strong, charismatic leader—the glue for Pakistani nationalism and Pakistan's democratic growth—died one year after the country's independence. No one can ever know how different the history of Pakistan could have been if the Quaid-e-Azam had lived for a decade, cultivating a successor generation of Pakistani democratic leaders and containing the growth of the military and intelligence agencies as political institutions. No one knows what a decade of strong democratic leadership from the Father of Pakistan would have meant to containing religious extremism, which he had confronted but which continued to rear its head throughout the ensuing sixty years of Pakistani history. No one can ever know, but certainly Pakistan would be a very different place today if Mohammad Ali Jinnah had lived. And if he were living today, he would doubtless be appalled by the perversion of the political system that he shaped by modern dictators and demagogues.

In any case, India certainly got off to a more democratic start than Pakistan, which allowed for democratic roots to take hold. India adopted its Constitution in 1949 and held its first national general elections two years later. Pakistan adopted its Constitution a full decade after independence, and only two years later a military coup d'état brought Pakistan constitutionally back to square one. Electorally, the comparison was even more disparate. While India proceeded with fair and free elections in 1949, Pakistan would not conduct transparent and free elections until 1970, a full generation after India.

An unfortunate confluence of circumstances stood in the way of democratic governance and democratic development in Pakistan. Islam was not the reason, although the politicization and manipulation of Islam by a series of authoritarian Pakistani regimes certainly contributed to recurrent crises of dictatorship in the country.

At the time of Partition, the new nation of Pakistan created a Constituent Assembly to write a constitution and establish a governing structure. The work of the Constituent Assembly manifested major problems of national integration that would erupt later in the nation's history. The Constitution was delayed due to strong disagreements over the different needs of East and West Pakistan. One major disagreement was over language. Those in the West wanted Urdu as the official langue of Pakistan, while those in the East wanted Bengali as the official language. After much violence over this issue, a compromise was reached that established both languages as the official languages of the new nation.

In September 1948 Mohammad Ali Jinnah died, and Liaquat Ali Khan, who was prime minister at the time, became the de facto leader of Pakistan. Liaquat Khan was the first prime minister of Pakistan, while the nation's founder, Mohammad Ali Jinnah, held the title of governor-general. The governor-general was the head of state, and the prime minister was the head of government. The 1956 Constitution changed the name of the head of state from governor-general to president. When Jinnah died, he was succeeded by Khwaja Nazimuddin as governor-general, but due to Liaquat's higher standing in the country and the weak and unestablished system of Pakistani governance, Liaquat was the effective ruler of the country. Liaquat Ali Khan was assassinated in 1951.

Following Liaquat's death there were two weak prime ministers: Muhammad Ali Bogra (1953–1955) and Chaudhry Muhammad Ali (1955–1956). Ghulam Muhammad was governor-general, but in 1955 Major General Iskander Mirza took the office for himself.

In February 1956 the new Constitution was finally adopted. The Constituent Assembly became the interim Parliament. General Mirza was elected Pakistan's first president by the Constituent Assembly. The first two years of Mirza's term saw great political instability in the country, with two successive prime ministers who served for only a matter of months. In October 1958 General Mirza started what would become a cyclical affair in Pakistani politics by declaring martial law. He announced that Pakistan must adopt a new constitution. Mirza's new cabinet included twelve members, but only four were Bengalis from East Pakistan, reflecting the early tension between the two wings. Almost immediately, Ayub Khan, the head of the military, staged a bloodless coup and exiled Mirza.

In 1960 Ayub introduced "basic democracies" that provided for localities to elect local representatives to represent them in government. These "basic democracies" did not have the power of an actual parliament, but they affirmed Ayub's presidency in a problematic referendum.

In 1962 a new constitution was promulgated for Pakistan. It banned political parties, which Ayub claimed corrupted politics. Ayub was forced almost immediately to back down from this position and allow for functioning political parties within the constitutional structure. The Constitution also set up an indirect election for the position of president. It also removed the word "Islamic" from the name of the country. An advisory committee had warned Ayub not to tamper with "Islamic" in the nation's name (the Islamic Republic of Pakistan), but Ayub refused to listen. His stubbornness backfired, and he put "Islam" back into the Constitution in 1963.

Ayub Khan believed that economic development would ultimately lead to democracy, or so he said. He actually assumed the position of chairman of the Planning Commission, which planned national economic policy, and elevated the vice chairman of the Planning Commission to a cabinet position. He claimed economic progress, with GDP growing at a 5.5 percent rate. However, the so-called fruits of this development did not trickle down to the poor masses. Instead he concentrated the wealth of the nation into twenty-two families, including his own, who controlled Pakistan's wealth.

During the 1960s, international academics argued that the new states needed "authoritarian leaders," under whose control economic development would take place. Economic development would, in theory, create a middle class, which ultimately could be the springboard for democratic development.

Such academics failed to realize that dictators disempower people and breed discontent. The dictators then crush first the press, to stop coverage of their repression, and then the judiciary, to stop justice for its victims. The very building blocks of democracy—political parties, media, and the judiciary—are the casualties of authoritarianism. Thus dictatorship fuels instability by creating a delusion of order to camouflage the great disorder below the surface.

Moreover, inevitably, without a free press, an independent judiciary, or a balance of power between alternating political parties, dictatorship breeds a lack of accountability. Corruption, nepotism, and greed spread. Monopolies are created, and artificial shortages drive up prices while providing quick profits to favorites. The result is a growing gap between the rich and the poor, while economic figures on GDP growth (often attained through reckless borrowing) seem buoyant.

In 1965, Fatima Jinnah, the sister of the Quaid-e-Azam, challenged Ayub Khan in the presidential elections. She traveled around the country, attracting very large crowds, charging that Ayub was a dictator, and demanding a return to democracy. Although she was defeated, her challenge exposed Ayub's vulnerability and the weakness of his popular support. The massive popular support she received on the streets showed that the people of Pakistan, "the real people," were sick of false promises of democracy and ready for a real leader to lead them forward. My father was close to Fatima Jinnah, in deference to our family relations to the entire Jinnah family. When he visited Karachi, he would call on her to pay his respects. Soon after her loss in the presidential elections, she died mysteriously. Some said she had been strangled to death by General Ayub's goons. A year after the presidential elections of 1965, my father resigned from the cabinet of President Ayub Khan.

Ayub Khan was a moderate leader, and in 1958 my father, who was then thirty years old, had joined his cabinet as South Asia's youngest cabinet minister. Educated and articulate, he brought with him the vast respect our family enjoyed in the province of Sindh due to my grandfather's movement to separate Sindh and to build the Sukkur Barrage. Moreover, as prime minister of Junagadh state, my grandfather had also been instrumental in advising that the Muslim ruler accede to Pakistan.

Although Ayub Khan's military dictatorship had started out with a promising note of social reforms and was initially popular, the gloss soon began to wear off. My father, like the rest of the country, was growing increasingly disillusioned with General Ayub. The early reformer had given way to a strongman who brooked no dissent and seemed to favor his family and friends over everyone else in the country.

National events then began to collide with international ones. Relations with India had been deteriorating all through 1964, and two events in particular set the relationship on fire. First, an important Muslim relic, what was thought to be a hair from the Prophet Mohammad, was stolen from a mosque in Srinagar in Indian-occupied Kashmir. Many Pakistanis charged that the Indian government was responsible and was trying to provoke a confrontation with Pakistan. In May 1964 Indian prime minister Nehru died and his successor, Lal Bahadur Shastri, did not have Nehru's charisma or popular support. He could not make the concessions necessary to jump-start a rapprochement with Pakistan. Later in 1964, the Indian-controlled Jammu and Kashmir's constitutional status changed when the disputed territory was brought more into line with the rest of the territories of the Indian Union. Pakistan believed this action had been ordered by Delhi as part of a plan to integrate Kashmir into India, demonstrating that the Indians intended to violate U.N. Security Council Resolution 38, mandating a plebiscite for the people of Jammu and Kashmir to determine their future.

Small-scale border skirmishes began between India and Pakistan in the area of Rann of Kutch. It turned into heavier fighting, which was ultimately stopped by British mediation. It resulted in both sides accepting the border status quo, which at the time was seen as a victory for Pakistan. This victory, compounding India's loss in the 1962 Sino-Indian War, convinced some Pakistanis that India's power was weakening. Tensions between India and Pakistan soon erupted and spilled over into war on September 6, 1965. I remember those days vividly. The country began building trenches, and sirens blared to warn us of incoming bombs. We heard the sounds of war as the planes zoomed overhead. The 1965 war was perhaps the finest hour for the men of the Pakistani armed forces. They heroically laid down their lives to protect Pakistan, tying bombs to their bodies and lying down in fields to stop invading tanks and prevent Lahore from capture. Young pilots were killed defending the skies. The entire nation was united in the defense of the country.

The international community intervened, and a cease-fire was signed on September 22, 1965. In January 1966 the Tashkent Declaration formally ended the war and essentially brought everything back to the prewar status. The Tashkent Declaration was seen by many in Pakistan as a capitulation to India. This very unpopular war did not sit well with many Pakistanis, who blamed Ayub for having given up too much too soon and embarrassing the nation. My father resigned, but President Ayub asked him not to leave the cabinet at such a critical time.

Six months later, in the month of June, my mother held a birthday party for me. There were balloons and buntings, cakes and pastries, and an atmosphere of celebration. I was excitedly opening my birthday presents and was thrilled that my mother had given me my first high-heeled shoes as I was now a teenager. Suddenly my father entered and the mood of celebration turned serious. We knew from the looks my parents exchanged that something important had developed.

My father took us all to a separate room and told us that he had once again resigned from President Ayub's cabinet. Despite Ayub's request to him to stay on, he said that six months had passed since the signing of the Tashkent Declaration and he could no longer stay. The Tashkent Treaty was the tipping point in my father's growing discontent with Ayub's rule. My father was thirty-six years old when he turned his back on the corridors of power and took on a mighty dictator.

My father resigned his position as foreign minister. He was determined to begin a grassroots political movement to take Pakistan from authoritarian rule to democracy. He created the Pakistan Peoples Party in November 1967, bringing together a coalition of civil society, social classes, and intellectuals with grassroots politicians. The students and youths in particular were drawn to my father's charisma, oratory, nationalism, and promise of freedom. The downtrodden social classes, the weak and the poor, responded to his call for an economically egalitarian society. Wherever he went, my father was mobbed. His speeches, which had mesmerized young diplomats at the United Nations, now became a clarion call for the people of Pakistan to gather under the banner of the tricolored PPP flag to fight for their fundamental human right to determine their own destiny.

Anti-Ayub demonstrations led by the PPP swept Pakistan throughout 1968 and into 1969. Ayub tried to mollify the protestors by freeing political prisoners, including Sheikh Mujibur Rahman, the most popular leader in East Pakistan. But the tide had turned against Ayub, and the power of the people was demonstrated to be a potent political force in Pakistan. The protests also established the bona fides of the Pakistan Peoples Party as a mass, grassroots, effective political organization.

As antigovernment riots spread across the country, my father began a hunger strike in Larkana. Soon lawyers, journalists, students, women, intellectuals, and labor and political activists joined him in going on hunger strikes across the nation.

Against the background of the escalating demonstration of people's power, General Ayub did what generals do best: he conducted a coup d'état against his own Parliament and government by declaring martial law for the second time. But the anger of the people did not die down. It increased. Ayub Khan refused to lift martial law and turn over the power under the provisions of his hand-drawn Constitution. Instead, on March 25, 1968, he stepped down, handing power to his second in command in the military, General Yahya Khan. The national and provincial assemblies were dissolved, the Constitution was abrogated, and political activities were banned. The artificially constructed Ayub "decade of development" collapsed into ruins, and Pakistan now had a second martial law administrator.

The martial law continued until November 1969, when General Yahya announced that he would hold direct elections the following October. Conceding to the opposition's demands that parliamentary representation be determined by the size of the population, Yahya did away with representational parity between East and West Pakistan in the National Assembly. Given the population distribution, this meant a quantum leap in representation to East Pakistan. In March 1970 Yahya issued the Legal Framework Order (LFO). Under the terms of the LFO, the new Assembly would be both a Constituent Assembly and a Parliament.

It would have 120 days in which to draw up a constitution. Many feared that the term limit was imposed to set the new Parliament up for failure. If the new Parliament failed to write a constitution within 120 days, the dictator had an excuse to say "Democracy has failed" and go back to dictatorship. If the Constitution were bulldozed by some of the parties, the dictator could reject it on the grounds that it was "against the unity and integrity of Pakistan."

Soon reports circulated that a journalist close to the intelligence agencies had been sent to Dacca, the capital of East Pakistan, to help the leading party there draft a constitution that would be rejected by the army in particular and Punjab in general on the grounds that it was a prescription for the breakup of Pakistan.

Pakistan was a key ally of the free world at the time. It was a partner in the military alliances CENTO and SEATO. The breakup of Pakistan would clearly open the way for the Soviet Union to expand its influence through disintegrating West Pakistan to the warm waters of the Indian Ocean through Sindh and Baluchistan.

This was clearly the "security nightmare" General Yahya planned to script to scare the Pakistani Army and the West into acquiescing to his continued dictatorship. The LFO established a 313-member National Assembly with an outright majority of seats going to the Bengalis in the East.

General Yahya and his security services never expected my father's newly formed party to do well. When Yahya met my father he told him this, pointing out that he would be able to win only Larkana and would lose the rest, including Thatta, a historical city near Karachi.

Thatta was famous for its archaeological finds as well as for a sweet drink, Thatta's sherbet, made from rose essence, almonds, and flakes of real silver, a unique recipe passed down from ancient times by word of mouth. My father said he would fight from Thatta himself, which he did and won. Thatta has remained a PPP stronghold ever since. Thatta proved how wrong the general's calculations were when he, in overconfidence, held the only fair elections conducted by a military dictator.

Free elections for the new National Assembly and the five new provincial assemblies concluded in December 1970. With Mr. Bashani's party boycotting the elections, it was a clean sweep for Sheikh Mujib and the Awami League in East Pakistan. The Awami League won

160 of the 162 seats allocated to East Pakistan in the National Assembly. Mujib's win had a lot to do with the pent-up East Pakistani grievances against West Pakistan, whose elite had exploited its resources and insulted its people, calling them "descendants of Dravidians who are short and dark compared to West Pakistanis, who are tall and fair" in the textbooks taught in school.

The Pakistan Peoples Party, led by my father, swept to victory in West Pakistan, winning an overwhelming share of the vote with 81 seats, primarily from Sindh and Punjab, on the slogan *"Roti, Kapra aur Makan"* ("Bread, Clothing and Shelter"). Political giants toppled before unknown supporters of my father who voted "for Bhutto," and it was said that "a lamppost with the PPP symbol would be voted for because he was Bhutto's candidate." This first set of true multiparty democratic elections was a big step forward for the development of Pakistani democracy. The jubilation was, sadly, short-lived. The Awami League's six-point constitutional proposal frightened the daylights out of the Pakistan security establishment, just as Yahya had planned. It threatened West Pakistan with disintegration. The newly created state of Pakistan was confronted with an ugly reality striking at the core of its security, stability, and unity.

Mujib insisted that since he had the majority in the National Assembly he would reject the views of the remaining federating units of Pakistan and impose his unilateral constitution on the country. This was contrary to the spirit of constitution making. A constitution is drawn up as an agreement setting the terms under which federating units voluntarily agree to live together. The PPP argued that a constitution could not ignore representatives from the rest of the country, where the Awami League had been totally rejected, failing to win even a single seat.

On January 3, 1971, Mujib once again rejected the right of other federating units to have a say in the framing of the Constitution. He said he would form a constitution based on his six points: (1) The Constitution should construct a federal state with supremacy of parliamentary legislature (in other words, Bengali domination of West

Pakistan in the formulation of all laws and resources to pay West Pakistan back for its domination of East Pakistan). (2) The national legislature should deal only with foreign affairs and defense. All other rights should be states' rights. (3) Two currencies should be introduced, one for East Pakistan and one for West Pakistan. (4) The federal government should have no taxing power. That power should be reserved for the provincial governments (in other words, there was no money to maintain an army and no money to conduct either a foreign or defense policy). (5) The two wings should have two separate accounts for foreign exchange. (6) East Pakistan should be allowed to maintain its own separate militia force (interpreted to mean that East Pakistan would have its own army).

Separate armies, separate currencies, separate central bank accounts, and no federal income to administer a federal government: Mujib's six points were akin to signing the dismissal orders of the entire civil service and military command. The state of Pakistan would be constitutionally dismantled through the Awami League's six points, letting the cat free among the pigeons. Neither the military nor the West nor the people of West Pakistan could countenance the breakup of a country for which so many had sacrificed their lives and suffered hardships just twenty-three years earlier.

Yahya's script was playing out perfectly. "Pakistan in danger" was a cry that he could exploit to scuttle the whole process. He could declare that "the country is not ready for democracy." Meanwhile, he had to go through the motions. So in mid-January, he traveled to Dhaka to see Mujib to work out an agreement. On January 17, he traveled to Larkana, the ancestral home of my family, to meet with my father. My father asked Yahya to either delay the holding of the Constituent Assembly or lift the 120-day limit for drawing up a constitution, to enable legislators to arrive at a consensus. Zulfikar Ali Bhutto continued to work for a constitution that would reflect the wishes of all five federating units of Pakistan. But Yahya was adamant. He would do neither—neither lift the 120-day ban for arriving at a constitution nor postpone the session to provide more time to thrash out a consensus. However, he said he was ready to consider it if Mujib agreed. My father traveled to Dhaka on January 27, hoping to persuade Mujib, but the meeting bore no fruits. Mujib insisted that he would show no flexibility on the six points, on the date scheduled for the Constituent Assembly, or on the 120-day limit to pass the new constitution.

At that point Yahya announced that the National Assembly would convene on March 3. My father represented the federating units in West Pakistan. For them Mujib's constitutional prescription was a recipe for the disintegration of Pakistan. As leader of West Pakistan my father concluded that he had two bad choices: He could go to Dacca, attend the Constituent Assembly, and acquiesce to an imposed six-point constitution, thus legitimizing it. Or he could boycott the session to visibly demonstrate the PPP's opposition to a unilateral constitution imposed on the nation by Sheikh Mujib. He declared that he and his supporters would not attend the constituent session of the National Assembly. He would not be party to the disintegration of all Pakistan through the dismantling of the federal state structure. He wanted consensus on the future constitution worked out before the Assembly's meeting while conceding that Sheikh Mujib had every right to form a government on his own once the Constituent Assembly session was over. There was little point in showing up for a session of the National Assembly, where the only point on the agenda would be to endorse Mujib's six-point constitution. Still Mujib remained adamant. He was ready to go it alone. He played into Yahya Khan's hands, surrounded by advisors, some of whom had been infiltrated by security agencies. On March 1, General Yahya postponed the convening of the National Assembly session.

Mujib had been preparing for this day of confrontation. Two days later, on March 3, he called for a statewide strike that paralyzed all of East Pakistan. Military troops and protestors clashed in what would be the beginning of months of violence in the province. Yahya backed down on March 6, summoning the National Assembly for March 25. Although Yahya gave in to Mujib, Mujib was not satisfied. Carried away by the successful strike call, he now said he would not participate in the National Assembly session unless his demands were met. These included: (1) "a judicial inquiry into the loss of life caused by military shootings since 1 March"; (2) "the immediate withdrawal of martial law"; (3) "the return of the troops to their barracks"; and (4) "the immediate transfer of power to elected representatives before the Assembly would meet."

A March 10 telegram from my father to Mujib requesting talks was rejected by Mujib. The Awami League had established a parallel government in East Pakistan. On March 5, General Yahya traveled to Dhaka to meet Mujib amid heavy security. My father went, too, on March 21, but the meetings failed. Mujib had, for all practical purposes, issued a unilateral declaration of independence. He had armed militias at his command. He controlled the country. And Bengalis no longer wanted anything to do with Pakistan. Either there would be a constitutional breakup of all of Pakistan, including West Pakistan, through the six-point constitution, or East Pakistan would secede.

The East Pakistanis then went on a rampage. The militias, the local police, and the East Pakistan Rifles began killing non-Bengali administrators. An "X" was marked on the house of non-Bengalis, marking them for death. The families of Punjabi officers were specially marked for death. A reign of terror was let loose. Blood flowed on the streets as the Pakistan flag was taken down and the Bengali flag was put up. The Biharis, seen as the collaborators of the Pakistani Army, were marked for death along with Punjabis.

No one had gauged the anger and hatred that had built up in the Bengali mind against Pakistan itself. Now it spilled out in a terrifying burst of violence and bloodletting.

On March 25, General Yahya ordered a military crackdown in a bid to forcibly keep East Pakistan part of Pakistan. As the crackdown began, students at Dacca University, in the capital, reacted. The army fired back, killing students and enraging Bengalis even further.

The uprising was intense. The people, the police, and the paramilitary forces had all risen up against what they called "the Punjabi

Army." Even as the army tried to quell the uprising, units of the army revolted against General Headquarters in faraway Rawalpindi.

Major Zia ur-Rahman (who would go on to become president of Bangladesh and his wife prime minister) was the first to mutiny. With the military defections adding to defections that had taken place in the police and paramilitary forces, the armed resistance for the formation of the independent state of Bangladesh was established.

As a brutal military crackdown began to crush the insurgents, Bengali women were raped and their male children slaughtered. Violence was met with counterviolence in a horrifying sequence of unfolding events. A million refugees fled from their homes across the border to India. Sheikh Mujib had been arrested and taken to a prison in West Pakistan. Yahya regretted not killing him.

In mid-April, a Bangladesh government in exile was formed in Calcutta, India, indicating the role the Indian government would play in the conflict. As the conflict spread, Yahya decided to put a docile government into place. Those who remained faithful to the Awami League were disqualified. Those who switched sides were accepted. By-elections were announced for 78 of 160 Awami League seats in the hope of forming a new parliament and government. But then the military turned for support to the religious parties, which had been defeated in the general elections. The military helped the religious parties organize death squads to further intimidate the Bengali population, thus cementing ties between the Pakistani establishment and the religious parties now based in Bangladesh.

The long, hot summer months were marked by state violence versus guerrilla violence. East Pakistan's economy was devastated. Rail lines were cut, trade was ruined, and war seemed imminent. In November, the Indian Army began a series of small attacks in East Pakistan and, starting on November 21, began amassing forces on the East Pakistani border. On December 3, Indian forces attacked en masse, starting the 1971 Indo-Pakistani War.

Within two weeks, Pakistani forces were routed. There was fear that West Pakistan, where India was making territorial advances, would fall, too. My father, as the elected leader of West Pakistan, had flown to the United Nations to plead for a cease-fire resolution, but international opinion was not with Pakistan. Everyone was waiting for the fall of Dacca. Unfortunately, it came too quickly, before pressure could be built up to force the hand of the international community. My father did succeed, though, in convincing Washington to save West Pakistan. President Nixon ordered the Seventh Fleet to Pakistan in the famous "tilt to Pakistan." Without that famous "tilt" all of Pakistan would have been threatened in 1971. On December 16, in a humiliating ceremony at Dacca Race Course in East Pakistan, General Yahya's generals surrendered to Indian generals. It was the darkest night in our country's young history. Not only had we lost our country, but we had been humiliated. Some of our jawans (soldiers) refused to accept the surrender and bravely fought to death. But the war was over. During the course of this two-week war, Pakistan lost half its navy, half its army, and one-quarter of its air force. More than 90,000 Pakistani soldiers were taken as prisoners of war, and, most important, Pakistan lost East Pakistan, which emerged as the new state of Bangladesh. Even as the nation tasted the bitter ashes of defeat, the generals did not want to end the dictatorship. However, the young officers refused to accept the dictatorship. When Yahya went to address them, he was hooted down. Realizing that the discipline of the Pakistan armed forces was under severe threat, even as the full danger to West Pakistan had not receded, the senior generals went to Yahya Khan and asked him to step aside and allow a civilian government to be formed based on the verdict and mandate given in the last elections.

As my father flew back from New York, the outraged people of Pakistan added their own pressure on the military brass. A massive public meeting at Rawalpindi's famed Liaquat Bagh was called to demand the resignation of General Yahya Khan and a transfer of power to the leader with the popular mandate, my father, Zulfikar Ali Bhutto.

On December 20, four days later, junior officers of the Pakistani

Army hooted Yahya Khan down, forcing him to resign. The army generals, realizing the enormity of the crisis caused by the massive embarrassment of the loss of half the country's territory, with Indian troops threatening West Pakistan and 90,000 prisoners of war in Indian camps, turned to West Pakistan's majority leader to take over. Zulfikar Ali Bhutto became the leader of the country. There was no constitution, so he was sworn in as chief martial law administrator and president of Pakistan under the Legal Framework Order. Within months of taking over, President Bhutto ended martial law and secured the approval from Parliament of an interim constitution. In 1973, Pakistan's first and only unanimously passed Constitution, drafted by elected representatives from all its provinces, was written and implemented.

It is more than a daughter's pride that causes me to evaluate the period of Zulfikar Ali Bhutto's leadership of Pakistan as a renaissance era for my country. It was a difficult time, for from the very outset (as would later be twice true for me), the military tried to destabilize his government. Although they were back in their barracks, the generals were plotting to take over again. My father faced two coup attempts even as he tried to save the country. The generals refused to accept any responsibility for the military loss to India but rather tried to deflect responsibility onto my father, claiming that his inability to reach an accommodation with the Awami League had been the catalyst for the war and the loss of East Pakistan. The military also organized and encouraged its allies in the religious parties to undermine Prime Minister Bhutto's leadership and destabilize his government. It was the beginning of a clear pattern of military cooperation with religious parties begun during the operation to suppress Bengali separation and continuing to this day.

I was present when my father negotiated the Simla Accord with Prime Minister Indira Gandhi in 1972 to return Pakistani territory lost in the western wing, repatriate the 90,000 prisoners of war, and save 5,000 people threatened with trial by a war crimes tribunal as well as seek a peaceful resolution of disputes between India and Pakistan. Actually, it was quite a triumph for Prime Minister Bhutto because he was dealing with an extremely weak hand yet managed to negotiate peace without having to recognize Bangladesh or accept a no-war clause with India. The Simla Accord also called for resumption of trade, overflights, and communications between the two states. It established the Line of Control in Kashmir (a de facto border), made possible the immediate return of lost territory to West Pakistan, and facilitated the return later of 90,000 Pakistani prisoners of war.

My father promoted an alternative to the religious and extremist parties. It is fashionable to describe this type of politics as "secular" although that does not fully connote its thrust. "Secular" in Urdu means "atheist." My father and the PPP believed that Parliament and the will of the people should be sovereign. They disagreed with the notion that religious scholars should determine the laws of the land or that the country should become a theocracy. They thought that the domination of the religious clergy, given the diverse schools of interpretation in Islam, would lay the field open to sectarian war.

My father's slogan was *"Roti, Kapra aur Makan"* ("Bread, Clothing and Shelter"), a call for economic and social development that continues to be the heart of the program of today's Pakistan Peoples Party. Rather than campaign on the rhetoric of Islam, he campaigned for the people and against the civil-military-religious establishment that had ruled Pakistan since its inception, and certainly after the death of the Quaid-e-Azam. Unfortunately, because of the traumatic breakup of Pakistan, which had created a state of insecurity and vulnerability, the bilateral relationship continued to cause defense spending to make up an inordinate percentage of the Pakistani government budget.

Because of the qualified secular nature of his party, campaign, and program, the religious parties, led by the Jamaat-i-Islami, were against him from the beginning, campaigning against him in 1970 and burning effigies of him in the streets when he took power in December. The army, which was still licking its wounds from two disastrous defeats by India in less than a decade, continued to align more closely with the religious parties. When Pakistan's surrender in Dacca was televised, Jamaat-i-Islami led protests against my father because they claimed he was trying to embarrass the army even though it had been the army and its Islamist allies that had gotten Pakistan into both of these national disasters. After the loss of Bangladesh, the Jamaat-i-Islami accused Prime Minister Bhutto of plotting to get rid of East Pakistan so he could consolidate his power in West Pakistan.

Not surprising, one of my father's first goals was to restore a proper relationship between civilian rule and the military. He aimed to make the Pakistani Army a professional defense force but remove it as a political force in our nation. Some said he was trying to weaken the army. That is not true. He was trying to reestablish civilian control over the army, which he thought was integral to democratic governance. To further these goals, he fixed the term of the army chief of staff at three years. He included in the 1973 Constitution a mandatory military officer oath prohibiting their involvement in politics. He also removed twenty-nine politically active senior-level military leaders from their posts within the first four months of his government's tenure.

As Ayub's cabinet minister in the sixties, my father had begun nuclear research. The first nuclear agreement was signed for the establishment of KANUPP in Karachi with Canadian support and international safeguards. In 1972 my father started research on Pakistan's nuclear program, declaring it was for peaceful purposes of acquiring environmentally clean nuclear power and reducing Pakistan's import bill for petroleum-based energy plants. Critics pointed out that peaceful nuclear products could be upgraded to weapons-grade nuclear material. Meanwhile, India was rapidly proceeding with its nuclear program; it would test a device in 1974, which it ironically dubbed "the smiling Buddha." When this happened, my father wanted to try to restore parity, a balance of force with India, which three years earlier had invaded our nation. He wanted to achieve what those in the West had been calling "mutually assured destruction," or MAD, as a deterrent to having any sort of nuclear exchange.

In 1972 Zulfikar Ali Bhutto implemented his commitments in the PPP Manifesto to nationalize the ten basic industries of Pakistan, including banking and shipping. He instituted a dramatic educational reform system that mandated compulsory and free education for all Pakistani children under the age of thirteen. My father began the process of practicing what he preached for the people of Pakistan— *"Roti, Kapra aur Makan."* He liberated women and minorities from their second-class position. He appointed the first woman governor of Pakistan, opened the subordinate judiciary, the police force, civil administration, and diplomatic corps to women, and set up a quota to ensure that minorities got jobs in key government institutions. He did away with separate electorates, giving minorities joint electorates in which they could vote for or against any candidate in every constituency of Pakistan. This empowered minorities. His labor reforms gave workers security of tenure, the right to be active in trade unions, bonuses, and other allowances. He introduced land reforms and broke the back of feudalism. The land nationalized was given to poor peasants. My father surrendered 40,000 acres of land, and other family members also surrendered their landholdings along with the rest of the country. Now each person could not hold more than 150 to 200 acres, a far cry from tens of thousands. It was the biggest shake-up of the social system. It made my father both permanent friends and permanent foes.

In August 1973, the new Constitution of Pakistan was accepted. It established a parliamentary system with a president who would serve as the honorific head of state, modeled after the British monarchy; a bicameral legislature with fundamental power resting in the popularly elected National Assembly; and an independent judiciary. It was a model democratic governing framework that paved the way for both economic and political development in Pakistan for the next four years.

In March 1976 my father elevated General Mohammad Zia-ul-Haq to army chief of staff, jumping over several more senior officers. It would prove to be a tragic mistake. Although Zia was personally religious, my father did not see this as a problem. One could be a pious person, he felt, without letting religion interfere in the politics of the country. My father did not know of General Zia's connections with Maulana Maudoodi of the Jamaat-i-Islami. Later General Zia would make the works of Maulana Maudoodi compulsory reading in the armed forces. The professional armed forces changed into one influenced by the politics of religion.

.

National elections took place in Pakistan in 1977, resulting in a resounding victory for the Pakistan Peoples Party, which took 155 of the 200 contested seats. An opposition alliance under the banner of the Pakistan National Alliance won 45 seats. PNA agitators led demonstrations around the country, alleging election fraud, which resulted—quite deliberately in many people's view—in violence and chaos. The protests were hijacked by the Islamists, who began demanding Sharia law in Pakistan.

On July 5, 1977, I was awakened by my mother while sleeping in the prime minister's house in Rawalpindi. She informed me that the army had taken over Pakistan in a coup d'état. Although the dictator General Zia publicly pledged elections within three months, there were no elections in ninety days, or in ninety months for that matter. Zia-ul-Haq, by lies and actions, set the cause of Pakistani democracy back by more than a generation. He would go on to charge my father with a baseless crime and, in the face of worldwide outrage, execute him on April 4, 1979, in what leading jurists referred to as a judicial murder.

On the day my father was arrested, I changed from a girl to a woman. He would guide me over the next two years, cautioning me to remain focused and committed and never bitter. On the day he was murdered, I understood that my life was to be Pakistan, and I accepted the mantle of leadership of my father's legacy and my father's party.

When Zia overthrew the elected democratic government of Pakistan, the religious parties celebrated in the streets and exchanged sweets as if it were the festive holiday of Eid. General Zia soon changed the motto of the army to *"Iman, taqwa, jihad fi sabil Allah"* ("Faith, piety and jihad for the sake of God"), which was an omen of things to come. The insidious axis between elements of the military and militant Islamists was cemented. It would rule Pakistan with an iron fist for a decade, and the effects of the damnable alliance would ripple across Pakistani society for generations and ultimately spill its venom into making my nation the petri dish of international terrorism in the new millennium. General Zia is "often identified as the person most responsible for turning Pakistan into a global center for political Islam." Zia not only attained power as a result of a mosque-military alliance but worked to strengthen that dubious coalition over the next decade.

Zia-ul-Haq claimed that Pakistan had been created in the name of Islam and thus he would impose an "Islamic system" on the country. Zia's military rule proved to be far more brutal and military-centered than Ayub Khan's government had been. He established military courts and had pro-democracy supporters lashed, tortured, and imprisoned. Zia put the military in the spotlight, relying on them and appointing retired military officers to important civilian positions. He transformed the Directorate for Inter-Services Intelligence into a tool of repression. He also introduced the concept of political Islamicization—the manipulation of Islam to consolidate political power.

The world condemned Zia's military dictatorship, which he dressed up as "Islamicization," especially the Hudood Ordinances, which were widely seen as abusive to women and did not differentiate between rape and adultery. The Islamist parties in Pakistan, however, reveled in their sudden legitimacy within the power structure. They supported Zia until his death and provided him the smoke screen of "Islamic" legitimacy for his tyrannical rule. Zia proved to be a loyal friend to his extremist allies, even changing the textbooks in schools to ones whitewashing their negative role during the Pakistan independence movement and the breakup of East Pakistan. The books used in the curriculum "supported military rule in Pakistan, inculcated hatred for Hindus, glorified wars, and distorted the pre-1947 history of the area constituting Pakistan." In colleges and schools of higher education, Zia purged secular professors and substituted them with supporters and members of Jamaat-i-Islami.

On top of all this, Zia's military junta was notorious for human rights abuses, including shooting protestors dead and hanging political activists. He was repeatedly cited and criticized by international human rights organizations such as Amnesty International. My party and family were particular targets of Zia's brutality. My mother and I still wear the mental and physical scars of the abuse we suffered in class C prisons on the orders of the Pakistani dictator (my beloved brother Gogi, Shah Nawaz, would be murdered in France in 1985 in a crime I always will believe bore the unmistakable stamp of Zia and his Secret Service henchmen). Each time Zia postponed elections, the choke collar around political parties and the civil society tightened further. When he announced a ban on all political meetings, the prescribed punishment for violating the ban was ten lashes and twenty-five years of rigorous imprisonment. He also adopted and enforced draconian press censorship. Every newspaper was censored before publication, often appearing with blank spaces where news items had been scissored out by the censors. The press was banned from printing the Bhutto name. All films and photos of the Bhutto period were burnt. Arrested journalists were handcuffed and chained. Zia also banned student and labor unions. He used the goons of the Jamaat-i-Islami to gun down progressive students in universities. Jamaat-i-Islami was given carte blanche to appoint university professors and intelligence officers from its student wing. Officers related to members of a moderate political party were weeded out of the army or simply not recruited.

In 1984, in a pitiful attempt to supply some degree of democratic legitimacy to his military junta, Zia organized a so-called plebiscite on his leadership. The language of the plebiscite was worded in such a brazen, distorted way as to make a vote against Zia a vote against the religion of Islam. He campaigned through the government-controlled media, yet no one was allowed to campaign for a "no" vote. Jamaat-i-Islami actively campaigned for a "yes" vote, arguing that the people of Pakistan could not vote against Islamicization. Zia claimed victory in this farcical referendum despite the fact that only a handful of Pakistani voters participated.

In 1985 Zia called for elections to the National and Provincial Assemblies but banned the participation of political parties. Because the elections were to be non-party-based, the Movement for the Restoration of Democracy (MRD)—an umbrella coalition led by the PPP—decided to boycott the elections. I will always think that the Movement's decision to boycott these elections, as unfair and unfree as they were, was a tactical mistake. A political party, in any democratic or even quasi-democratic system, needs continual testing. It requires that its organizational machinery be kept in the highest state of readiness. Even if the 1985 elections were to be rigged, going through the exercise would have kept the gears of the political machinery oiled. Confronting the déjà vu of staged elections again at the end of 2007, it makes me revisit this critical strategic decision in political organization: boycott, and the dictator doesn't need to rig the election to get a pliant parliament; or participate, thus keeping the machinery oiled and through popular support forcing him to rig and thereby deny the new Parliament legitimacy.

After the non-party-based elections were completed, Mohammad Khan Junejo was nominated as prime minister. Yet Zia's civilian government was in fact nothing more than a sad charade. Though martial law was officially lifted after the elections, it, for all intents and purposes, remained intact. General Zia continued as head of the army, or "Army Chief of Staff" until the day he died. He had armed the presidency with overwhelming power. He didn't actually need the formality of martial law to continue his tyrannical rule. He had the power to dismiss governments, dismiss the prime minister, dis miss the National Assembly and provincial assemblies, and appoint provincial governors, the heads of all the armed forces, and the Joint Chiefs of Staff. With that in his arsenal, who needed the formality of martial law to keep a dictatorship intact? The end of martial law was nothing more than a publicity stunt for the West.

I had suffered in Zia's prisons and under house arrest for nearly six years after my father's democratic government was overthrown. In 1984 I was released for medical treatment in the West. During my time in prison, an ear infection had left me virtually deaf in one ear. I suppose that Zia, who already had killed my father and my brother, did not want to face the international uproar of having yet more Bhutto blood on his hands. In any case, I was finally released from prison and allowed to travel to London for treatment. He would later confess to one of my friends, in a message meant for me, that letting me live had been the gravest mistake of his life.

Whether I had ever wished it or not, I was the leader of not only the Pakistan Peoples Party but the entire democratic opposition in Pakistan. I was a symbol of democracy, and that responsibility weighed heavily on my shoulders. I used my time after I was released from prison to organize the democratic opposition from London and to lobby for Pakistani democracy throughout Europe and in the United States. I walked the halls of the House of Commons and pounded the paths between the U.S. House of Representatives and the Senate. I cultivated the civil society network all over the world and met almost continuously with the international press that cared about the future of South Asia. But I reached a point where I recognized that my fight for democracy in Pakistan could not continue to be run from abroad. If I were to win the battle against dictatorship in Pakistan, it was time for me to return home and confront the Zia-ISI-jihadist axis, irrespective of the dangers that were entailed. On April 10, 1986, one million Pakistanis in Lahore greeted me and expressed their support for democracy.

Taking advantage of General Zia's fear of a Bhutto return to power,

Prime Minister Junejo began to assert himself. He invited me to an All Parties Conference to discuss the Geneva Accords setting a time-table for a Soviet withdrawal from Afghanistan. The meeting took place in the State Guest House in the lane running up to the Army Chief of Staff House in Rawalpindi. Zia was furious. A tiny parliamentary opposition had also emerged, and it began to reach out to me. Meanwhile the public and the press were responding to me. Zia could not stomach it.

In May 1988, Zia-ul-Haq sacked Prime Minister Junejo, dismissed the National Assembly and provincial assemblies, and appointed himself the caretaker president. He wanted a parliament whose members would not meet me and a prime minister who would not invite me for political consultations. His intelligence had reported that I was expecting a child. Two days afterward, his governor's newspaper reported that Zia would call for fresh elections because I was pregnant and "could not campaign." Zia called for general elections. Fearing a PPP win, Zia declared that the elections would be held on a "nonparty" basis.

We will never know whether Zia-ul-Haq would have succeeded in rigging the elections against the PPP, or even if elections would have ever occurred. He died in a plane crash on August 17, 1988, returning from a military base in Bahawalpur. Democracy would be given another chance in Pakistan, but the damage that Zia had done to democratic development in Pakistan was incalculable: "Zia left behind not only a political process distorted by the Eighth Amendment, which enabled his successors to dismiss elected prime ministers with impunity, but an atmosphere of bigotry, fanaticism, and distorted values." The president of the Senate, Ghulam Ishaq Khan, took over as acting president pursuant to the Constitution.

In looking back on the Zia regime, international factors contributed to the strengthening of the Zia dictatorship, just as they would help to empower the Musharraf dictatorship a generation later. The war that raged in Afghanistan between the Soviet Union and the mujahideen gave Zia the freedom to do almost whatever he wished domestically. The United States, fixated on defeating and humiliating the Soviets in Afghanistan, embraced Zia and the ISI as surrogates; the United States' attention was riveted exclusively on Afghanistan, disregarding the war's impact on internal political factors in Pakistan. Thus, out of miscalculation or indifference, America did not seem aware of the consequences of Zia's Islamicization or of his strategy of aligning the Pakistani military, the intelligence agencies, and the jihadist parties into a long-term political coalition to reshape Pakistani politics and Pakistan's role within the community of Islamic states. Starting with the Soviet invasion of Afghanistan on Christmas Eve of 1979, Zia started increasing the size of the ISI to help support the mujahideen in their battle against the Soviets and their battle to radicalize the influence of religious factions within Afghanistan. Over the course of the war, the ISI helped channel at least $8 billion from the United States to the mujahideen and untold billions from the Persian Gulf states and Gulf religious organizations to the mujahideen.

Zia needed this money from the United States, and he used it to secure and consolidate his power within Pakistan. He had killed my father and abandoned his call for elections. By supporting the mujahideen with U.S. money, he dramatically strengthened his support in Pakistan with both the military and the Islamic parties. Zia leveraged the war and played the money card very well. In January 1980 he rejected a $400 million aid package from President Carter as "peanuts." With the election of President Ronald Reagan, the political and monetary relationship between the United States and Pakistan changed dramatically. Reagan offered Zia an aid package of $3.2 billion, eight times what Carter had offered. The United States also began sending into Afghanistan, through Pakistan, surface-to-air shoulder-fired Stinger missiles. Each dollar and each advanced weapons system sent to Afghanistan was used by Zia to consolidate his own political power within the military and to empower the radical political parties within Pakistan. The U.S. policy achieved its short-term goals but with horrific long-term consequences, not only for Pakistan but for the entire world.

The adage of "follow the money" in the 1980s would be repeated twenty years later as the United States made an almost identical financial commitment of overt and covert aid to the Pakistani military, to obtain access to landlocked Afghanistan through Pakistan for supplies and flights. Jacobabad Air Base was given to the United States as a base. However, as before, there was no accountability about where the funds and equipment provided to Islamabad actually wound up.

The alliance between elements of the Pakistani military and religious political parties began before the Zia years but reached its zenith under his dictatorship. In the 1960s, Jamaat-i-Islami began receiving substantial funding from Saudi Arabia. During the Afghan war against the Soviets, Jamaat-i-Islami was used to contact its counterparts in other Muslim countries to recruit for the Afghan mujahideen.

The leaders the Jamaat-i-Islami signaled out for support, the ISI supported, too. The Afghan resistance leaders had to draw close to the ISI and Jamaat-i-Islami to get the military and financial support necessary to mount an effective campaign. Those leaders included Professor Abdul Rassul Sayyaf and Gulbuddin Hekmatyar.

While Pakistan was serving the interests of and acting as a surrogate for the United States throughout the 1980s in Afghanistan, it was also following its own self-perceived interest of trying to set up a friendly Afghan government to act as a buffer and give Pakistan "strategic depth" against India. The concept of "strategic depth" was a theory with almost universal backing within the officer corps of the Pakistani Army, which added to the zeal of Pakistan's military efforts with the United States in Afghanistan. Pakistan was not just acting as a surrogate for a superpower in the Cold War. Its actions in Afghanistan synchronized with Islamabad's strategic and political goals of gaining influence in Kabul, which had traditionally been pro-India while claiming a neutral status.

In April 1988 the Geneva Accord arranged for the Soviet with-drawal from Afghanistan. The Soviet rout was complete. The Americans had achieved their goal to a greater degree than they had dreamed possible. The military, political, and economic hemorrhage of the nine-year battle in Afghanistan had depleted the Soviet Union of the ability to survive as an unified state and of the resources it needed to compete internationally as a superpower. The Soviet Union splintered into individual republics and withdrew its political and military presence from Central and Eastern Europe. The Cold War, which many had referred to as "World War III," was over, and the West had won. In doing so, however, it may have sown the seeds for the successor generation of world conflict for the new era, what some have actually called "World War IV."

Zia's death and the end of the Afghan war coincided; both threatened the ISI-Jamaat lock on political power in Pakistan. This power arrangement was cemented under the formation of the Pakistan Muslim League. The party was composed of political leaders groomed and funded by the ISI, which functioned in tandem with the Jamaat-i-Islami. For good measure, it co-opted other religious parties. These included Jamiat Ulema-e-Pakistan, led by Maulana Noorani. The ISI splintered Jamiat Ulema-e-Islam into two. The JUI group led by Maulana Sami-ul-Haq was co-opted into the ruling alliance. The JUI group led by 245 members joined the opposition MRD. The ruling alliance was flush with money, opportunism, and stalwarts who had become rich overnight through either political madrassas (religious factories) or "permit industries" (inefficient industrial units run on written-off loans from public sector banks). For the 1988 elections, the forces of tyranny refused to let go and developed a strategy to attempt to retain power. The intelligence agencies assembled a coalition of seven Islamist parties around the Pakistan Muslim League to run against the Pakistan Peoples Party in the National Assembly elections. This ISI-created political chimera was called the Islami Jamhoori Ittehad (IJI) (Islamic Democratic Alliance).

Jamaat-i-Islami, which had supported Fatima Jinnah as a presidential candidate in the 1960s, now proclaimed to the world that Islam prohibited a woman from becoming prime minister of an Islamic state. The intelligence agencies poured hundreds of millions of rupees into the IJI campaign in a desperate attempt to hold on to the reins of power.

In 1988, the electoral deck was stacked against the PPP. The state-controlled media were clearly hostile to the forces of democracy. At the time the literacy level in Pakistan was 26 percent, and political parties were recognized by their symbols. Our party symbol was snatched from us, and we had to find a new one. Almost unlimited amounts of money poured in from the ISI to fund the IJI campaign.

The acting president unconstitutionally passed a decree changing the election laws to purge several PPP candidates from the ballot. He also declared that all voters must have a national identification card, although he knew full well that in the rural areas of great PPP strength, only a third of the men and a mere 5 percent of the women actually had these cards.

Nevertheless, we moved forward with our message of reconciliation and redirection of national priorities. We reached out to all democratic elements of our nation, to the business community, to the civil society, and even to moderate elements within the military, to assure the nation that our agenda was moderate democracy and that we would reject extremism in all of its ugly manifestations.

Clearly the institutional political environment was hostile. The election was hardly free and definitely not fair. Massive rigging took place in both the voting and the counting. Twenty-six national parliamentary seats were rigged, but so massive was the support for the PPP that political giants came tumbling down. I remember a friend of mine ringing me from America and asking how things were going electorally. I replied, "Humpty Dumpty sat on a wall. Humpty Dumpty had a great fall. All the king's horses and all the king's men couldn't put Humpty together again." He got angry and said, "Be serious." But the "king's" men did fall down. These included the all-powerful IJI leadership—spiritual leader Pir Pagara, Chairman

Ghulam Mustafa Jatoi, former prime minister Junejo, and my father's cousin Mumtaz Bhutto, who had long ago broken from the PPP—among others. I still remember, as I campaigned, how the crowds shouted, "Bhutto family, hero, hero, the rest zero, zero." Unbelievably, with all the rigging and manipulation we won an outright majority. All of the king's horses and all of the king's men failed to put ISI's Humpty Dumpty in control again. Despite the disenfranchisement of potentially half the Pakistani electorate, the Pakistan Peoples Party won a majority of the seats in the National Assembly. The PPP swept the province of Sindh, not conceding a single seat to the IJI, and won a clear majority in the Punjab. We emerged as the only party to win seats in all four provinces of the country, winning 108 seats, including members elected from tribal and minority seats in the National Assembly, compared to the ISI-backed IJI's 54. (Our figure went up to 122 when we won 14 of the 20 women seats. Women were indirectly elected by members of Parliament. I won a direct seat.) Unfortunately, the ISI succeeded in blocking a landslide win giving us a two-thirds majority. Such a majority was necessary to make constitutional changes. Without those changes, the stage was set for the dysfunctional democracy of the 1990s.

Having failed to stop the PPP from winning a parliamentary majority, the ISI now began plotting to dismember the PPP. Makhdoom Amin Fahim of the PPP was met by General Hamid Gul, the head of the ISI. (He has publicly admitted that as the head of ISI he formed the IJI.) Makhdoom was asked to defect and told that if he could bring ten members of Parliament with him, he would be made prime minister of Pakistan. He refused.

Despite the two-to-one victory of the PPP over the IJI for the National Assembly, President Ghulam Ishaq Khan defied the parliamentary norm of inviting the leader of the winning party in a parliamentary democracy to form a government. Instead he called different leaders to see if he could cobble together a coalition government with the support of small parties and independents. But since the PPP had an outright majority, he could not. After dillydallying with coalition building for fifteen days, he was forced to call on me to form the government. However, during this critical fifteen-day lapse, the enemies of democracy contrived an outcome in the powerful Punjabi Provincial Assembly that manipulated the vote to allow for an ISI-IJI majority (Nawaz Sharif, a Zia protégé with Islamist leanings, was made the chief minister). Despite all the odds, on December 2, 1988, I was sworn in as the democratically elected prime minister of Pakistan, the first woman in history elected to head an Islamic state. This was the moment of democracy, the moment to honor all who had come before, who had given their lives or been tortured and lashed and exiled while fighting for freedom.

In my first days in office, I honored my commitments to the people of Pakistan. I practiced what I had been preaching. I moved to send a clear signal to Pakistan and to the community of nations both substantively and symbolically. I returned democratic governance to the people of Pakistan. I freed all political prisoners. I restored free, open, and uncensored print and electronic media. I allowed CNN—which was then the only global electronic medium—in and allowed newsprint to be freely imported into Pakistan. I removed constraints and conditions to the free operation of NGOs, including women's and human rights groups, so that Pakistan could truly achieve a functioning civil society. I opened up the state media, for the first time in the history of Pakistan, to regular, frequent, and uncensored access by the political opposition. I lifted the ban on student and labor unions that had been imposed by the Zia military junta. My government granted an amnesty to all political prisoners and exiles, including Baloach and Pakhtun nationalist leaders who had gone into exile. We began the separation of the judiciary from the executive, which I completed in my second term. We introduced the computerization of identity cards to document our citizens and provide the basis of a fair electoral process. It would take more than a decade before the program was completed. We introduced microcredit and protected minorities' rights.

What we did in the private sector was unprecedented, even revolutionary. My government made Pakistan the first country in South Asia and the Middle East to introduce privatization of public sector power units. We deregulated the financial institutions. We decentralized the economy and freed it from bureaucratic red tape. We funded the electrification of 40,000 villages in Pakistan. We built more than 18,000 primary and secondary schools. Exports increased by 25 percent. Foreign investment quadrupled.

In foreign policy, we made broad overtures even to those who had been our adversaries—and, of course, to those who had stood by us in the past. I am particularly proud of our work with Indian prime minister Rajiv Gandhi, building on the progress in Pakistan-Indian relations that our parents had established in the Simla Accord. Rajiv and I negotiated a remarkable treaty committing our nations not to attack each other's nuclear facilities. This was the first nuclear confidence-building treaty between Pakistan and India. We established a hotline between the General Headquarters of both our countries. We reached a draft agreement on the Siachen Glacier for withdrawal of troops by both sides to Kargil as well as on mutual reduction of troops. The ISI tried to undermine the budding relationship between two young, moderate, post-Partition leaders who were willing to "think out of the box" to break the stalemate between our two nations. They accused me of being an "Indian agent." The Muslim League and Jamaat-i-Islami took up the chorus. However, time proved the sincerity of our initiative. Both the Muslim League government that followed, as well as the military under Musharraf, would build on these initiatives. The Muslim League would call it "bus diplomacy" and General Musharraf's regime "composite dialogue."

We also made tremendous progress in rebuilding the relationship with the United States. Many expected U.S. aid to Pakistan to all but disappear after the Soviet withdrawal from Afghanistan on February 15, 1989. Instead, our team in Islamabad and Washington worked to get the White House and Congress to greatly increase aid to Pakistan, making the country the third-largest recipient of foreign assistance from the United States, after Israel and Egypt (whose aid level formula was fixed as part of the Camp David Accords). We negotiated a nuclear confidence-building measure with the United States, making "no export of nuclear technology" part of our nuclear doctrine. We also decided not to put together a nuclear device unless the country's security was threatened.

We supported the establishment of an Afghan interim government to pave the way for representatives of the Afghan mujahideen to form the government as and when power was transferred from President Najibullah's government to that of the true representatives of the Afghan people. This was headed by the moderate president Sibghatullah Mujadidi.

The PPP government made dramatic reforms in women's rights. I appointed several women to my cabinet and established a Ministry of Women's Development. We created women's studies programs in universities. We established a Women's Development Bank to give credit only to enterprising women. We created institutions to help in training women in family planning, nutritional counseling, child care, and birth control. And we legalized and encouraged women's participation in international sports, which had been banned in the years of the Zia military dictatorship. It was a solid start in a society where Islam had been exploited to repress the position of women in society for a bitter generation.

From the very outset of my tenure, indeed even before I took office, there was a concerted attempt, once again, by the hard-liners in the ISI, led by the trio of General Hamid Gul, Brigadier Imtiaz Ahmed, and Major Amir to undermine the democratically elected government's ability to govern. The installation of Nawaz Sharif as chief minister of Punjab and his declaration that he would defy my writ and make me the "prime minister of Islamabad" was a major destabilizing force in building national consensus. I was attacked as un-Islamic and a tool of Washington. But there was another scheme concocted against me that was more personal and hurtful: there was a concerted campaign against my character and the reputation of my family that was planned in the electoral cells of the ISI even before our electoral victory. Long before it became more common as a political tool in the United States and Europe, the ISI invented the "politics of personal destruction," a deliberate and methodological program to sully my name and suggest that my government was corrupt. My businessman husband was a key target of this conspiracy (dubbed with the pejorative "Mr. Ten Percent" title a month before I took the oath of office), a strategy that I doubt would have been used against a male prime minister. Even my friends abroad became vilified in the Urdu press as "Hindo-Zionist conspirators."

It was not just me whom the intelligence community and its extremist collaborators were trying to destroy. They understood full well that after a decade of dictatorship, and after a series of military dictators throughout Pakistani history, the efficacy of democracy was on the line, the demonstration that under democracy government can successfully address the quality of life of the people. The ISI-jihadist axis knew that by undermining me they were undermining the chance of democracy to take root and succeed in Pakistan and undermining the growth of the moderate political center that marginalizes extremists. The forces of the past had a great deal at stake in trying to convince people that our government had failed to function successfully.

In 1989, with money provided by Osama bin Laden, a no-confidence motion was moved against my government. Efforts were made to buy the loyalty of members of my parliamentary group. They were offered $1 million each. Some of those approached were appalled and took me into their confidence. I kept them as my "Trojan horses" in the ISI-IJI camp so that the numbers against me would be miscalculated.

I used another group to videotape Brigadier Imtiaz Ahmed asking my members to defect because "the army" did not want me. And I worked on members of the opposition who had known my father or were disgruntled with the IJI.

I had the numbers and, much to the shock and dismay of the president, the army chief, the ISI, and Osama bin Laden, the Pakistan

Muslim League's no-confidence proposal was roundly defeated on the floor of Parliament.

·

I tried to take control of the ISI by removing General Gul and replacing him with General Shamshur Rahman Kallu. But General Gul got President Ishaq Khan and General Mirza Aslam Baig, the army chief of staff, to authorize the transfer of the ISI's duties to Military Intelligence (MI). MI was then headed by Brigadier (later General) Asad Durrani. He was General Gul's deputy, as Gul had headed MI before moving on to the ISI. While the ISI's ability to destabilize the government was neutralized, the military security campaign continued under the aegis of the MI.

Under the political cover of the world's attention being fixated on Saddam Hussein's invasion of Kuwait, the forces of extremism used the onerous provisions of the Eighth Amendment of the Constitution (which allowed the president to dismiss an elected government and call election within ninety days) to bring down my government on August 6, 1990. My husband was arrested soon thereafter. Both of us, as well as our followers, faced trumped-up charges before special tribunals. A witch hunt was launched to cripple us politically.

The head of MI, General Durrani, now took over the ISI. The ISI went for the PPP, and my family, with a vengeance. Vicious advertisements were placed in the newspapers. My family and I were accused of corruption, arrested, and vilified. A special cell was set up to rig the elections. It was then that I accused the ISI of operating independently for political purposes, exploiting the name of the army. I said it had turned the army against me, that my government had been destabilized by those who had fought the Afghan jihad, who had supported Zia's tyranny, his so-called Islamicization, and saw the PPP and democracy as a threat to its aims.

The ensuing Pakistani elections were a charade. Our party workers were thrown in jail. Following the collapse of the Mehran Bank, the head of the ISI, General Durrani, would admit that the ISI had invested 60 million rupees obtained from the private Mehran Bank to ensure a PPP defeat. I have always wondered how much more must have been taken from other banks to fund the 1990 election campaign against me.

And on September 30, 1990, the U.S. government formally cut off all foreign aid to Pakistan, citing the Pressler Amendment on nuclear proliferation. During my tenure, I was able not only to keep aid to Pakistan stable but actually to substantially increase the flow, including military assistance, even though the surrogate war across the board in Afghanistan was over. Yet just fifty-eight days after a coup against me generated by the military and the ISI, all military assistance to Pakistan was terminated. It was a great irony. The electoral fraud reached its denouement on October 24, 1990. The ISI rigging restricted the PPP, the largest political party in Pakistan and the only national party with strength in all four of Pakistan's provinces, to an absurdly low 44 seats in the National Assembly. Statisticians determined that up to 13 percent of the national turnout for the ISI-backed Pakistan Muslim League (the principal component of the IJI) was fraudulent, approximately 6.5 million votes. The electoral sham gave the PML, now led by Nawaz Sharif, an absolute majority in the National Assembly and control of each of the four provincial assemblies around the nation. I was elected Leader of the Opposition.

The Nawaz administration almost immediately sought to reverse many of the social programs that I had instituted. Press censorship was reinstated. Student unions were once again banned. Access to the media for the opposition was blocked. The Nawaz budget shifted emphasis and funds from the social sector to the military, in order to appease the IJI's mentors who had engineered its electoral victory. The budget for education programs was especially hard hit. And quite painfully to me personally, many of the reforms that I had instituted for women and girls in Pakistan were undone, including closing down the women's health and population control centers to appease the regime's supporters in religious parties.

But being the Leader of the Opposition in a parliamentary structure did allow me an important mechanism to challenge the dictates of the new regime. In the great parliamentary model of Britain, as Leader of the Opposition I could question and debate government officials and even the prime minister, and I used that power to try to preserve the social, economic, and political accomplishments of my government and to moderate the government's attempts to turn Pakistan into a theocratic state. My party blocked Nawaz's attempts to pass an "Islamicization" bill that could only have fueled sectarianism and dictatorship in the country. (The Afghan interim government also collapsed, and as Kabul fell, civil war and bloodshed continued in Afghanistan.)

Brigadier Imtiaz Ahmed, who had been videotaped conspiring to destabilize the elected government and retired in shame from the army, was rehired as head of the Intelligence Bureau. A dirty tricks campaign started against the PPP. An airliner was hijacked as part of a conspiracy to slander the PPP but the plot was foiled by foreign authorities before it could succeed.

A police van escorting me was blown up in a rocket launcher terrorist attack to kill me. I narrowly escaped. All those in the van were killed. It was a difficult and challenging task for me, but the PPP did its best to reinforce the idea that Pakistan had the functions of a democracy. The more I exercised my role as Leader of the Opposition, the more I felt that the dream of democracy could be reinforced in my homeland.

Nawaz's first term in office coincided with a series of international terrorist attacks. Blasts took place in Bombay, and the World Trade Center in New York was attacked for the first time. Pakistan was soon on the brink of being declared a terrorist state.

I believe that the age of international terrorist war actually coincided with the suspension of democracy in Pakistan. Acts of terrorism—although not fully appreciated until the second attack on the World Trade Center in New York in 2001—became the centerpiece, the cause célèbre of a violent movement that seeks to provoke war between Islam and the West. This is what Professor Samuel Huntington of Harvard University had predicted in his *The Clash of Civilizations and the Remaking of World Order,* which I discuss in chapter 5.

In January 1993, General Asif Nawaz, the moderate army chief of staff, died under mysterious circumstances. General Nawaz's death threw Pakistan into political turmoil and undermined Prime Minister Nawaz Sharif's support within the military. As a response to the shifting political tide, Nawaz Sharif went on television and accused President Ghulam Ishaq Khan of undermining his government, triggering the president to use the powers of the Eighth Amendment to dismiss the prime minister and dissolve the National and Provincial Assemblies. A technocrat from the World Bank, Moeenuddin Quereshi, was appointed head of a caretaker government, and elections were scheduled.

During the election campaign that followed Nawaz Sharif's ouster, extremists conspired to block my election as prime minister. The intelligence was no longer satisfied with trying to keep me out of power by funding the opposition. They wanted a permanent solution to their Bhutto problem. I was a major obstacle to their ultimate dream of the establishment of an extremist bloc of nations bringing the Muslim countries together in a borderless society. In the fall of 1993, my assassination was ordered and the chosen assassin was a Pakistani with ties to the ISI during the Afghan war. He was named Ramzi Yusef. Yusef failed twice to kill me during the election campaign in the fall of 1993. He had earlier planned and executed the first attack on the World Trade Center on February 26. The man who supplied Yusef with weapons to assassinate me was Khalid Sheikh Mohammad. After 9/11 he was identified as a mastermind of Al Qaeda and arrested from the house of a Jamaat-i-Islami supporter.

On October 24, 1993, the Pakistan Peoples Party won the elections for the National Assembly. As the election results came in, it was clear that the PPP was winning. But then the results stopped. It was sickeningly clear to me that the cause of the results slowing was ballot tampering.

We rang everyone we knew in an attempt to build pressure to announce the results expeditiously. In the early hours I retired and did not even know the next morning whether the results were in. The results came after a delay of about twelve hours. We had won by the skin of our teeth. I rushed from my hometown of Larkana to Lahore, the capital of Punjab, to shore up our strength, which now depended on coalition partners.

As prime minister, I worked quickly to implement a social action program that addressed the long-neglected needs of the people of Pakistan, concentrating on education, health, housing, sanitation, infrastructure, and women's rights. In the first year of the new PPP government, Pakistan attracted four times more private sector international investment than the previous year. This was more in that one year than the nation of Pakistan had been able to attract in the past twenty years of its existence as an independent state. A heavy percentage of this new foreign investment went directly into the energy sector, into power generation, to jump-start the stalled Pakistani economy. We modernized the stock exchange and computerized the State Bank. We trained 100,000 women to work in the towns and villages of Pakistan in health and family planning. We built 30,000 new primary and secondary schools, bringing the total new school construction during my two tenures to 48,000 schools. And convinced that the most effective way to expand child literacy was to have literate mothers, we instituted a remedial education program specifically targeting women.

We expanded digital and wireless communications and introduced computer education and software generation. We involved high school students as volunteers to ensure universal immunization of infants against polio, a disease that had been eliminated in most countries but not Pakistan. We literally doubled tax revenues and tripled the national growth rate. Pakistan became one of the ten emerging markets of the world. We were moving; we were modernizing; we were bringing Pakistan into the twenty-first century. But that very program of modernization made us even more of a threat to the extremists, who thrive on ignorance and poverty to achieve their suffocating political agenda.

It was not surprising that the extremists and authoritarians would view me and my government as major obstacles to their designs and goals. The extremists, Islamists, and intelligence agencies coalesced and mobilized to undermine our economic and social policies and discredit my government. They did not like my efforts to back the formation of a broad-based Afghan government with the help of the United Nations. They thrived on anarchy. My government, on the other hand, was restoring law and order and reforming the political madrassas. We ended the army operation in Karachi and brought peace to that great city, which had been taken hostage by terrorists. We cracked down on kidnappings for ransom, on the drug trade and illegal immigration. The nascent international terror cabal feared Pakistan when it was in my control. They saw my government, with its pro-people aims, as a threat to their ideological and theocratic agenda. And I stood in their way. I believe that there is at least some degree of causality in the fact that most of the major terrorist attacks in the world took place when my party and I were out of office, when they could operate without check or oversight. This includes both the 1993 and 2001 attacks on the World Trade Center, the Bombay blasts, the Indian Parliament attack, the attacks on the U.S. embassies in Africa and the U.S.S. *Cole,* the train attacks in Spain, and the subway attacks in London.

I believe that if my government had continued for its full five-year term, it would have been difficult for Osama bin Laden to set up base in Afghanistan in 1997, when he established Al Qaeda to openly recruit and train young men from all over the Muslim world. In 1998, two years after my overthrow, Al Qaeda declared war on the United States. Today Al Qaeda and the Taliban operate in parts of FATA and parts of the North-West Frontier Province of Pakistan with impunity, arrogance, and brutality. Democracy removes the oxygen from the air of the extremists. They understand that better than anyone else and deliberately target democratic forces throughout the Islamic world as a strategy for expanding their goal of an obscurantist empire, which they call "Islamic." Muslims such as myself reject the notion that Al Qaeda, the Taliban, or extremists are the real face of the Muslim masses.

My government and I were threats to the extremist forces that under General Zia's tyranny had coalesced together under an axis of the Afghan mujahideen and Pakistan's security services. The Afghan mujahideen coalition put together by the ISI of the 1980s has now broken into the Arab Al Qaeda, the Afghan Taliban, the political madrassas in Pakistan (and perhaps elsewhere), elements of the intelligence, military, and political groups. Pakistan's political grouping, which the military dictator of the eighties put together, subdivided into the Muslim League, Jamaat-i-Islami, the JUI (S), and militant groups such as Jaish-e-Mohammad and Lashkar-e-Tayyabba. In my second term I had good relations with my military chiefs, including General Wahid Kakar and General Jehangir Karamat.

I also initially had good relations with the ISI, when its chief was General Javed Ashraf Qazi. We worked together to save Pakistan from being declared a terrorist state. They supported the government by remaining neutral during the days when terrorist activity was at its height in Karachi. The military and intelligence were pleased that my government had withdrawn the army from Karachi, relying instead on the police and paramilitary troops to aid the government in restoring peace. They helped us reform the political madrassas. They helped us uncover an Islamist coup d'état led by Brigadier Billa Muntasir.

I had returned from Washington in 1995 with arrangements for $368 million in new military equipment. Under my leadership the army and the police began participating in international peacekeeping missions. It made them proud that the world honored them and

Pakistan as among the best in peacekeeping missions of all those in the world.

Unfortunately, things changed after General Mahmood Ahmed became head of MI and Generals Bakhtiar Rana and Chaudhry Shujaat Hussain took over the ISI and its internal wing. (General Mahmood later became the head of the ISI. He helped General Musharraf seize power in a coup. He was removed after 9/11.)

I suspected that the Zia elements were working on the chief justice of the Supreme Court. I went public on this, believing that the chief justice was trying to create a constitutional crisis in the belief that he would be made the interim prime minister after my overthrow. I complained to the army chief that General Mahmood was destabilizing my government.

The Zia elements also worked on President Leghari. The army chief tried to help the government. However, he was constrained by the constitutional power of the president over the army. The president had the power to dismiss both Parliament and the army chief.

The strategy of destabilization of my second PPP government brutally came home to haunt my family on September 20, 1996, when my brother Murtaza was gunned down in a police shoot-out in front of our family house. My father and two brothers were now dead. It was especially bitter for me because after three years of political disagreements, Murtaza and I had warmly reconciled only two months before. I was convinced that Murtaza had been killed as part of the conspiracy to overthrow my government. And to add salt to the terrible wound inflicted on my family, the intelligence began swirling rumors across Pakistan that somehow and some way and for some unknown reason, my husband was connected to my own brother's death. The rumors continued even after a judicial inquiry concluded, after my government's overthrow, that my husband had not been involved. In fact the judicial inquiry blamed other "high officials," but they were never investigated.

Using the American presidential election as a political distraction, my government was brought down on November 4, 1996, with authorities once again citing the traditional criteria of the Eighth Amendment, vague and unsubstantiated charges of "corruption and mismanagement." A National Defense and Security Council was created that formally institutionalized the Pakistani military's role in the political and policy process of the nation. The murder of my brother threw Pakistan into turmoil. It was an open secret that President Leghari was about to dismiss my government. Taking advantage of the turmoil, the Taliban marched into Kabul, taking over most of Afghanistan. The agreement reached with the help of the U.N. special envoy, which was scheduled to be signed on November 6, was scuttled with the overthrow of my government on the night of November 4, 1996. The hard-liners had won. They had "strategic depth" in Afghanistan. But that "strategic depth" would soon turn into a "strategic threat." With the rout of the Taliban in 2001, the ragtag elements of Taliban and Al Qaeda poured across the mountains of Tora Bora to take refuge in the tribal areas of Pakistan. Soon they would reorganize and begin expanding their influence in Pakistan. Simultaneously they attacked NATO troops in Afghanistan to destabilize the government of President Hamid Karzai.

Pakistan today is the most dangerous place in the world. Pakistan faces the threat of both Talibanization and Balkanization, which are gaining in strength. Back in 1997 the fraudulent elections were the coup de grâce to the takeover by the sympathizers of the extremists and militants. According to reports, the elections were rigged out of special election cells within the presidency. The journalist who disclosed this news was imprisoned, as was the computer expert who had put together the system. There was nothing subtle to the jiggering this time. Out of a 207-member parliament, the PPP, the largest political party in the four provinces of Pakistan, was allowed to win just 19 seats. I was, once again, the Leader of the Opposition, but I wasn't allowed to sit in the chair of the opposition leader. I was left to work from the back benches. My husband was once again behind bars. Special courts were established with handpicked judges, and special laws were passed. The new accountability chief operated from the prime minister's office.

The newly installed prime minister, Nawaz Sharif, proceeded to implement a reactionary social and political agenda. He attempted to have Sharia law incorporated into the Constitution of Pakistan. He praised the Taliban society as a model for Pakistan to follow. *The New York Times* said Nawaz's bill gave the prime minister, not the courts, the power to enforce religious edicts.

In May 1988, India tested five nuclear weapons, and despite the pleas of Washington and London, Pakistan followed suit within a fortnight with six underground tests of nuclear devices of our own. Strong sanctions followed suit, isolating Pakistan from the international community. Nawaz Sharif responded by freezing private foreign currency accounts. This created a panic in the market. The economy, already under strain from his profligate ways, went into a tailspin.

In October 1998, Nawaz Sharif chose General Pervez Musharraf as army chief of staff. It was a decision that would come back to haunt Nawaz the next year, and the world for almost a decade. In May 1999, the Kargil conflict brought Pakistan and India to the verge of nuclear war. Militants surreptitiously infiltrated Indian-held Kashmir in the Kargil region. Upon discovery of the infiltration, India counterattacked and mobilized 200,000 troops. The situation threatened to degenerate into full-scale war. By July, Nawaz had gotten himself into an intractable position and sought help from President Clinton to end the crisis. He traveled to Washington over the American Fourth of July holiday. According to Bruce Reidel, a national security official, President Clinton was "angry" with Nawaz Sharif, "complaining that Pakistan had promised but failed to bring Osama bin Laden to justice from Afghanistan" and that Mr. Sharif had allowed ISI "to work with the Taliban to foment terrorism." Nawaz was told that the United States expected Pakistan to withdraw from Kargil and defuse the crisis. Sharif returned to Islamabad weakened and embarrassed. He ordered a unilateral withdrawal. As no plan was worked out with the Indians for a peaceful withdrawal, hundreds lost their lives as they withdrew. The Indians shot them down as they regained the peaks. Others died of frost and cold. It was the most humiliating moment for the military since the fall of Dacca. It soured relations between Nawaz and Musharraf and a blame game started. Nawaz claimed he had been unaware of the Kargil plan. The military said he had known. It was just a matter of time as to who would move first against the other.

In October 1999, Nawaz tried dismissing Musharraf as army chief. The generals resisted. By seeking absolute power and dressing it up in Islamic garb, Nawaz had alienated the political parties and civil society. He found himself isolated.

He had removed his benefactor President Leghari, who by dismissing my government had paved the way for Nawaz's rise. His cabinet ministers participated in the mob attack on the Supreme Court of Pakistan. Judges were forced to flee the courtrooms as furniture was broken and a mob bused in by government charged into the Supreme Court. The media were hounded, including the country's largest press group, Jang Publications, while the editor of *The Friday Times* was threatened with sedition charges.

Musharraf seized control of the government in a military coup d'état on October 12, 1999. General Pervez Musharraf, in the modus operandi of Pakistani dictators of the past, announced with great fanfare a three-pronged plan for the restoration of Pakistani democracy: economic development, government accountability, and "true democracy." At the time the Musharraf regime was an international pariah. Then came the events of September 11, 2001. After war-gaming whether Pakistan could take on the United States and finding it couldn't, Musharraf chose to ally himself with the United States in the battle against terrorism. According to him, the U.S. official Richard Armitage threatened to "bomb Pakistan back to the Stone Age" if he didn't. Once Musharraf threw in his lot with Washington, the sanctions from the 1998 nuclear tests were lifted. Weeks later, U.S. Secretary of State Colin Powell delivered a $650 million aid package, the first installment of what is now a flow of $10 billion in overt (mostly military) assistance, and additional covert "black money" conservatively estimated at $5 billion. Additionally, the day after Powell traveled to Islamabad, the European Commission authorized a trade concession to Pakistan worth $1 billion. All of Pakistan's foreign debt was forgiven or restructured, enabling Musharraf to create an aura of prosperity for Pakistan's business and military elites. In fact for seven years, the growth rate remained under 6 percent, which the PPP had achieved without extended economic largesse.

In 2002, with the NATO war against the Taliban and Al Qaeda raging across the border in Afghanistan, Musharraf amended the Pakistani Constitution to increase presidential power ahead of scheduled parliamentary elections. Musharraf orchestrated a mock referendum, just as Zia had done eighteen years before, to declare himself president for five years. (Musharraf claimed a 60 percent turnout, while no international observers could find more than a 17 percent turnout, and probably much less.)

In the ensuing parliamentary elections, the leaders of both the PPP and the PML (N) were banned from contesting. I had been viewed as a threat to military dictatorship since the 1977 coup against my father. Now Nawaz Sharif was often out with the Pakistani security establishment because of his feud with Musharraf. However the Zia/ISI party the Muslim League, which Nawaz had led, continued in power under a new leader. This was Nawaz's number two, Chaudhry Shujaat.

Despite all the obstacles, the Pakistan Peoples Party won the largest number of votes. It was in a position to head the government either with Musharraf's allies or with the rest of the opposition. Fearing a PPP government, General Musharraf unilaterally postponed the Parliament session by edict. His henchman General Hussain Mehdi and Home Secretary Punjab Brigadier Ejaz Shah (whom Musharraf later promoted to chief of the Intelligence Bureau) broke up the PPP's parliamentary strength using a carrot-and-stick approach. Ten PPP members defected, and the ISI called all parliamentarians to its headquarters in Islamabad. There the parliamentarians were ordered to vote for Mir Zafarullah Khan Jamali, a former Zia cabinet minister, as prime minister of Pakistan. Jamali won with a one-vote majority made possible by the blessings of the ISI and the forced defection of ten PPP parliamentarians. The political attack on the PPP always seems to go back to the Zia forces that formed the Afghan mujahideen, who morphed into the militants who today threaten the integrity of Pakistan and peace in the larger world community.

For reasons best known to Musharraf, the general provided for the victory of an Islamist coalition called the MMA to take over the provincial governments of Baluchistan and the North-West Frontier Province. This outcome was curious since international observers labeled the election a fraud and the results manufactured. There is only one rational answer: General Musharraf wanted to exaggerate the so-called strength of the Islamist threat to Pakistan and paint his authoritarian regime as a bulwark against it. He jiggered the poll numbers in these two provinces to prove his point. He could and would use the results in Baluchistan and the NWFP as a scare tactic in the world community to convince them that he was the only thing that stood in the way of a nuclear-armed, fundamentalist, mullah-led Pakistan. The fact that the Islamist parties received only 11 percent of the vote, even in the hyped numbers of 2002, and had failed to poll at more than 5 percent in previous elections did not seem to register in the West. Musharraf was to be the fair-haired boy in the alliance against terror, and facts would not be allowed to confuse the conclusion. As in the 1980s, a Pakistani military dictator would once again play the West like a fiddle, using a short-term military objective to obscure and rationalize the consolidation of dictatorship in Pakistan. Just as Zia-ul-Haq repeatedly said he would have elections in Pakistan and never did, General Musharraf assured the country and the world that he would resign as army chief of staff. However, he kept postponing the date until domestic and international pressure coalesced to force him to do so in November 2007.

Though Musharraf quickly installed first Zafarullah Jamali and then Shaukat Aziz as prime minister in 2002, there was no doubt in Islamabad, Washington, or London who ruled Pakistan. The rubber-stamp National Assembly dutifully passed everything that General Musharraf sent before it with the support of Jamaat-i-Islami and other religious parties.

What is perhaps most remarkable about the Musharraf dictatorship is the disconnect between its rhetoric and its actions concerning the containment of extremism and the pursuit of Al Qaeda, the Taliban, and Osama bin Laden. Publicly it proclaims fidelity to the battle against terror and indispensability to NATO and America. Yet it simultaneously refuses to commit the political capital or human resources necessary to the fight. Indeed, under General Musharraf's regime, the defeated and demoralized Taliban have regrouped and reasserted themselves and now pose a serious threat to the takeover of Pakistan. General Pervez Musharraf's regime seems to be the only one on the planet at this time that has actually signed a peace treaty with the Taliban, ceding chunks of the tribal areas to the Taliban and Al Qaeda in what are called the Waziristan Accords.

As I write, the flag of the Taliban flies over parts of tribal territories. They intimidate the settled areas of the Frontier Province and force the closure of girls' schools, barbershops, and video stalls. They are training youth in paramilitary techniques. They are housing, arming, and equipping terrorists. By the end of 2007, the people of Swat, other areas of the Frontier Province, and FATA fell under their shadow. If neighboring Besham falls, the militants will be approximately a hundred miles from the capital city of Islamabad. The Taliban are slowly controlling larger parts of the country. Pakistan's north is a perfect demonstration of why dictatorship cannot defeat extremism.

But it is not just in the FATA and Frontier Provinces where the extremists are running amok and the government seems unable to contain them. The Lal Masjid (Red Mosque) complex siege in 2007 is another example. After the elections of 2002, the Ministry of Religious Affairs appointed Maulana Abdul Aziz the imam (or leader) of the Red Mosque. The government looked the other way as the militants built a political madrassa complex housing thousands of students in the heart of the capital city. The militants took over a children's library belonging to the government and, without building permits, constructed a fortified nest for militancy. When the brother of Maulana Aziz, Maulana Abdul Rashid Ghazi, was arrested for smuggling weapons into Islamabad, he was released. In early 2007 the followers of this madrassa complex began a series of activities to "purify" the capital of Islamabad, including raiding music shops, beating women on the street, and kidnapping ten Chinese nationals whom the leaders of the mosque had characterized as prostitutes. Bands of burqa-clad young women and girls roamed the streets of the capital, blocks away from the Parliament and presidency, hitting and terrorizing the public with sticks and verbal abuse. The government did nothing. The police stood by and watched. The militants thus became emboldened and empowered and became increasingly violent, forcing the government to finally take action. On July 3, 2007, government troops surrounded the mosque and laid siege to it. The leaders of the mosque threatened waves of suicide bombers. Eventually the mosque was raided and dozens were killed. If the Musharraf regime had stood up to the extremists in the mosque when they first took it over and terrorized the citizens of Islamabad, this would have been avoided. Cynics believed that the regime "set up" the Maulanas to scare Western public opinion in an election year with the collateral purpose of frightening it into acquiescing to the deferment of democracy. Were the Maulanas and their supporters set up as stool pigeons in a more complex game to hold on to power?

When world attention turned to the incident, the Supreme Court took *suo moto* notice of the Red Mosque incident. The Supreme Court ordered the government-owned children's library to be returned to Maulana Aziz and Ghazi's families, for the complex to be rebuilt, and for compensation to be paid to those affected. Two of the judges who provided relief to the Red Mosque Madrassa Complex militant group were retained by General Musharraf's regime when he reshuffled the Supreme Court later in the year.

General Musharraf made a fundamental political miscalculation when he summarily sacked the chief justice of the Supreme Court of Pakistan, Iftikhar Mohammad Chaudhry, on March 9, 2007. The legal community reacted, believing the chief justice had been removed because he could not be counted on to rubber-stamp General Musharraf's "reelection" by the same national assemblies fraudulently elected in 2002 and to approve his reelection as president while Musharraf retained the office of army chief of staff. The sacking of Pakistan's top judge provided the opposition parties and civil society with a stick with which to take on the powerful presidency. The first to protest was the Lahore High Court Bar Association, on March 12, 2007. Several parliamentarians of the PPP, including Senator Latif Khosa, Mahreen Raja, and the president of the bar, Ehsan Bhoon, were brutally beaten. The pictures of Senator Khosa with blood pouring from a head wound and the other injured advocates enraged the members of the bar associations, who called a protest meeting in the Lahore High Court. The police raided the High Court, forcing the judges to flee. This enraged the lawyers further.

Chief Justice Iftikhar was removed from office on charges of corruption. When the first hearing of the court case came up, the president of the Islamabad Bar Association, who is also a member of the PPP, made a protest call. This was joined by other political parties and civil society. The police reacted by raiding the Geo television station to stop it from broadcasting the protest. This angered the media. The chief justice's removal became the catalyst to trigger pent-up fury against General Musharraf's regime. Seeing the reinstatement of the chief justice as a setback to General Musharraf's attempt to consolidate power, postpone elections, and defer the return of democracy encouraged the PPP and others to join a movement for the reinstatement of Justice Iftikhar. The Supreme Court restored the chief justice in 2007.

However, General Musharraf went ahead with his presidential re-election scheme by the lame-duck National Assembly, the Senate, and the provincial assemblies on October 6, 2007. The Supreme Court, surprisingly, did not stay the election. Musharraf "won," taking 386 votes against 300 who did not vote for him. A switch of 44 votes would have resulted in his defeat.

.

I returned to Pakistan after eight years abroad on October 18, 2007, and was greeted in Karachi by crowds estimated by Sindhi press and party officials to be up to three million people. It was a moment I had dreamt of for so many years. I was overwhelmed by emotion as I touched the land of my birth and saw the love of the people. It was a love I returned with all my heart and soul. Politics started out as a duty for me. Over the years of pain, suffering, sacrifice, and separation, of young men and women tortured and killed, it had become an all-consuming passion.

When I returned, I did not know whether I would live or die. I was told by both the Musharraf regime and a foreign Muslim government that four suicide bomber squads would attempt to kill me. These included, the reports said, the squads sent by the Taliban warlord Baitullah Mahsud; Hamza bin Laden, a son of Osama bin Laden; Red Mosque militants; and a Karachi-based militant group. I said farewell to my children, husband, mother, staff, friends, and family not knowing whether I would ever see their faces again. I told my children, "Do not worry. Nothing will happen to me. God will protect me." I wanted to reassure them, but I also told them, "Remember: God gives life, and God takes life. I will be safe until my time is up." I wrote a letter to General Musharraf. I told him that if I was assassinated by the militants it would be due to the sympathizers of the militants in his regime, who I suspected wanted to eliminate me and remove the threat I posed to their grip on power. Deep in the heart of Karachi, the crowd celebrated people's power in joy and festivity. Suddenly the streetlights went dim and shut off.

The jammers we needed to prevent improvised explosive devices from going off stopped working after eight hours. Senator Rukhsana Zuberi went to the KESC, the power station, to try to get the lights switched on, which was necessary for our security to work effectively. The lights stayed off. Ten hours after my arrival, I went down to the compartment in the truck to take off my sandals. My foot had swollen from standing for so long, and the sandal was cutting into it. Now I was physically closer to street level.

Suddenly two powerful bombs went off in the space of forty-five seconds. In between a flamethrower was used that sent a blaze of orange fire into the sky. The truck rocked from side to side, sickeningly. Two dents in the side of the armored truck meant it had been hit directly. I looked outside. The dark night was bathed in an orange light, and under it dead bodies, crumpled bodies, lay scattered in the most horrific scene of bloodshed and carnage. The music had gone, the songs of celebration were silenced. There was only a deathly quiet, the quiet of the graveyard.

About thirty minutes before I went down to loosen my sandals in the compartment on the lower deck of the truck, I saw a man holding up a baby. The baby was dressed in PPP colors and was about one or two years old. The man gesticulated repeatedly to me to take the baby. I gesticulated to the crowd to make way for him. But when the crowd parted, the man would not come forward. Instead he would try to hand the baby to someone in the crowd. Worried that the baby would fall and be trampled upon or be lost, I would gesticulate no, you bring the baby to me. Finally he pointed to the security guard. I asked the security guard to let him up on the truck. However by the time he came to the truck, I was going to the compartment. Agha Siraj Durrani, a PPP parliamentarian, was watching the access to the truck by the stair. When the man tried to hand the baby up the truck, Agha Siraj told him to get lost. The man then went to a policemobile to the left of the truck, which refused to take the baby. The man moved to the policemobile in front of the first. A woman councilor of the PPP, Rukhsana Faisal Boloch, was on this mobile, as was a cameraman. As the man tried to hand the baby to the second policemobile, the first policemobile made an announcement: "Don't take the baby, don't take the baby, don't let the baby up on the truck." Both these policemobiles were exactly parallel to where I was sitting in the armored truck. We suspect the baby's clothes were lined with plastic explosives. As the man scuffled with the police in the van to hand the baby over, the first explosion took place on the van. Everyone in that van was killed, as were those around it. Human flesh, blood, and body pieces flew everywhere. The blood and gore rose up to the truck, sticking to the clothes, hair, and hands of the people I had been standing up with earlier. The van went on fire right next to the truck. Within fifty seconds, a fifteen-kilogram car bomb was detonated, throwing people up far and wide, penetrating them with pellets, shrapnel, and burning pieces of metal. According to some eyewitnesses, snipers began firing at this point. Had the jammers worked, the bombs could not have gone off. Had the lights been on, the car with the bomb would have been spotted.

Although I came out of the truck about eight minutes later, there seemed to be some chemical in the air. I suffered like others from both a perforated eardrum and a racking dry cough, the like of which I had never had before. Meanwhile, instead of rushing away, the Jaan Nisaar Benazir (my security guards) rushed toward the truck to secure me, taking the place of their brothers who had been killed. Less than an hour earlier, I had been talking to those in the front of my truck. We had been passing water to one another. I can still see their smiling faces, their eyes lighted up with the joy of the victorious reception. I shall see their faces till the day I die.

Dr. Zulfikar Mirza, who helped take the dead and wounded to the hospital, told me of the strange state of the bodies. The clothes of some were totally burned off. Others were clothed, but when one moved to pick up the body, it would melt and disintegrate. Many with pellet wounds subsequently died, making us suspect that the pellets had been soaked in poison.

When all the bodies had been counted, the number of those killed went up to 179 dead and nearly 600 wounded, some disabled for life. Yet the flame of hope burned so brightly that one mother told me, "I lost one son—I would sacrifice another to bring change." A young widow, married two months earlier, said her husband insisted, "I will take Bibi home safely and then return even if I have to give my life." Humera Alwani, a PPP parliamentarian, told me she had heard a chief minister say on television, "Benazir's procession will be over after midnight and she will never get to Bilawal House [my home]." The first bomb struck at 12:06 A.M. on the morning of October 19, 2007. Later, as I watched a video of the blast, I could hear the faint cry of *"Jeay Bhutto"*—"Long live Bhutto"—from the wounded as they lay dying in the streets.

Back at Bilawal House, our first thought was to arrange blood for the wounded and ensure they got to the hospital. Twenty of the dead around the truck could not be identified. They were so badly blown up that only body parts remained. Sixteen of them had heads separated from the bodies. We plan to bury them in the shadow of my father's final resting place at Garhi Khuda Bux Bhutto in my hometown of Larkana with honor and respect. Every year when the party gathers from the four corners of Pakistan to pay tribute to my father on the anniversary of his martyrdom, we will pray for the victims of Karsaz, the bridge near which the explosions took place in Karachi.

Later I was informed of a meeting that had taken place in Lahore where the bomb blasts were planned. According to this report, three men belonging to a rival political faction were hired for half a million dollars. They were, according to my sources, named Ejaz, Sajjad, and another whose name I forget. One of them died accidentally because he couldn't get away fast enough before the detonation. Presumably this was the one holding the baby. However, a bomb maker was needed for the bombs.

Enter Qari Saifullah Akhtar, a wanted terrorist who had tried to overthrow my second government. He had been extradited by the United Arab Emirates and was languishing in Karachi Central Jail. According to my second source, the officials present at Lahore had turned to Qari Saifullah Akhtar for help. His liaison with elements in the government, according to this source, was a radical who was asked to make the bombs, and asked for a *fatwa,* or edict, making it legitimate to oblige. He got the edict.

The bomb blasts took place in the army cantonment area in Karachi. When the army officials arrived on the scene, the wounded hooted them down. Rightly or wrongly, the perception in Pakistan is that the military is responsible for the rise of militancy and all the horrific consequences that it entails for Pakistan and its people. Militancy started with the Zia military dictatorship. Its heirs destabilized democracy until military dictatorship was once again imposed. The Musharraf military dictatorship, notwithstanding its public pronouncements, has presided over the mushrooming growth of militant groups and militant acts that have exacted a heavy human toll.

The crime scene was not quarantined. The evidence was contaminated. The government of Pakistan refused to allow forensic experts from the FBI and Scotland Yard to assist in the investigation, fueling speculation that it had something to hide. Then General Musharraf decided to suspend the Constitution. Although he called it "emergency rule," the suspension of the Constitution order was signed by him in his capacity as army chief of staff, thus establishing the action as martial law. He shut down independent electronic media and harshly censored the print media. He arrested thousands of lawyers, judges, human rights activists, and political party leaders. When my party gave a protest call for November 9, 2007, he arrested more than 5,000 leaders and workers of the PPP and put me under house arrest. When my party announced a Long March from November 13 to November 17 through central Punjab, he arrested me again, as well as 18,000 PPP supporters nationwide who were waiting to show the world that democracy still lives within Pakistan, even if it is under siege by a military dictator.

We had hoped that 2007 would be a year of progress, a year of transition to democracy leading to free and fair elections at the end of the year, and at times we believed that progress was actually being made. General Musharraf had become increasingly beleaguered despite his uniform and military infrastructural support. Political parties that he had desperately tried to suffocate had not only survived his onslaught but grown in strength. The civil society, and especially lawyers and judges, were growing increasingly independent. The press, and especially the private television media, allowed access to opposition leaders, unlike the state-controlled media. Consequently, their popularity increased rapidly. And the West seemed to be losing patience with the general's modus operandi against terrorism—"talking the talk" but not "walking the walk." In 2006, with an eye on the coming election, General Musharraf opened channels with both the PPP and the Nawaz faction of the Muslim League. He needed to demonstrate that he was broadening Pakistan's political base.

There had been an ongoing dialogue between the PPP and the Musharraf regime from the very outset of his rule. These negotiations failed every time because I insisted on a road map to democracy. With the failure of each round would come a new round of persecution. It was a mutual cycle of running around in circles. In 2000 I was asked to quit politics for ten years, in exchange for my husband's release and the dropping of unproven charges against us. I refused.

In 2002 I was rung up by my husband at my cousin Professor Ahmed Ispahani's house in California. High ISI officials were sitting with him. If I agreed to quit politics for ten years, he told me, he would be released and all persecution, for that is what the allegations were about, would end. I couldn't do it. I thought of my father in the squalid death cell and how he had refused to abandon the people of Pakistan at the cost of his life. In 2004 some of my friends intervened and my husband was released unconditionally. We hoped that this would help defuse the political tension in the country. It didn't. When my husband returned to Pakistan after visiting me, his reception was disrupted and he was repeatedly arrested and released. PPP supporters were also arrested, released, rearrested, their homes raided, their lives disrupted, their families traumatized.

The regime had released my husband, Asif, unconditionally and promised a level playing field to all political parties. It didn't happen. The local elections scheduled for later that year were fully rigged, and relations between the two sides soured.

When my husband visited the children and myself in Dubai in 2005, he suffered what one doctor described as a "near-catastrophic heart attack." It was a terrible period for us. Although doctors said Asif was not to be stressed, the regular pressure of judicial abuse and threats of ex parte action for a gravely ill man on whose behalf we could not answer was horrible. At times I wondered whether Asif would survive because his heart, blood pressure, and diabetes were in such bad shape. We were both hounded by court notices, appearances, mudslinging in the press, attacks on our character and on our reputation. I slept very little, spending an inordinate amount of time discussing legal issues and preparing material for the press to defend our reputations. This relentless pursuit, almost to punish us, went on for two and a half years before judicial matters went back to their normal beat of hearings. Asif had gone to New York for his treatment. I found myself flying to the United States every three weeks. I developed a problem climbing up and down stairs, which forced me to have a keyhole knee operation. When I look back at that period of my mother's illness, when she nearly died in 2003, and my husband's illness in 2005, when he nearly died, I don't know how I got through it all and dealt with party affairs, other opposition political parties, the diplomatic community, court cases, press, lectures, and our children. I worried about traveling in case anything happened to my mother or husband while I was away. I dreaded a phone call in case it was bad news related to them. I couldn't sleep at night, waking at any slight sound or imagining it in case something happened at night.

I had been lobbying internationally for a return to democracy and for my safe return to Pakistan, just as my party had been working on these issues in Pakistan. The question arose as to how there could be a smooth transition to democracy when General Musharraf was a key ally in the war against terrorism and the PPP, the most popular party, according to the elections of 2002, and he were at loggerheads. It seemed inevitable that a rapprochement was needed if Pakistan were to break the cycle of dictatorship feeding into the needs of an extraordinary security situation as the world confronted the forces of extremism.

For the PPP it was essential that General Musharraf retire as the army chief of staff. The PPP had consistently fought against military dictatorship, and therefore it was impossible for us to reconcile with a president who was also chief of the army. Yet for General Musharraf the army uniform was like a second skin. It was essential for the PPP that democracy be restored, and that could only occur through fair elections in which all political parties had a level playing field.

I refused to meet General Musharraf's emissaries, who insisted that as a precondition I opt out of the next general elections. Each offer was followed by a fresh bout of pressure in a ruthless abuse of the judicial system. Finally they dropped the precondition, but I did not want to meet them unless General Musharraf called me to assure me that they had his mandate.

I was in New York in August 2006 when General Musharraf called. He said that he wanted the moderate forces to work together. He asked for my support for a bill related to women's rights. I agreed, subject to a parliamentary committee's working out the details. It was agreed that a team would meet with me. This included the director general of the ISI, who has now become the strongest man in the army as its head, and the national security advisor. A step had been taken for confidence building between the two sides in August 2006. The passage of the women's bill gave momentum to the process of negotiations, although deep suspicions existed. To overcome this suspicion, I had suggested direct talks, believing that if General Musharraf met with me it would demonstrate concretely that he was willing to review his public posture that there was no place for me in national politics.

Throughout the process of dialogue, I kept London and Washington and a small group of PPP leaders briefed on the progress. As General Musharraf's biggest international supporters and key donors to Pakistan, the voices of London and Washington in support of democracy were essential.

In life small incidents are catalyst to bigger changes. The former mayor of Blackburn, England, Mr. Salis Kiani, was such a catalyst.

He hailed from the same constituency as British foreign secretary Jack Straw. At a time when I was persona non grata, he kept pushing Britain to lend support to a democratic Pakistan. He persuaded Jack Straw to meet with me to discuss the political situation in Pakistan. I used my time internationally to meet policy makers and make new friends who could help me put across the argument for a democratic Pakistan.

If there is a silver lining to the cloud of living away from home, it is in the people I met and the friends I made in those years. I met PPP leaders to discuss what our agenda should be in talks with General Musharraf. These issues in particular crept up: the uniform issue, fair elections, lifting of the ban on a twice-elected prime minister seeking office a third time, the balance of power between the president and prime minister, and an end to political persecution.

After several rounds of talks with General Musharraf's emissaries, a meeting was set up between General Musharraf and myself in Abu Dhabi, the capital of the United Arab Emirates. Even as the negotiations continued, so did the harassment, leaving me to conclude that the regime wanted to provoke me to pull out. So began a process, a dialogue that appeared at critical points to be moving the country forward to a democratic transition.

General Musharraf and I met in Abu Dhabi in January 2007. A helicopter was sent to pick me up, and I landed in the green gardens of the palace where the meeting was to take place. I thought of the time that I had earlier visited His Highness Sheikh Zayed bin Sultan Al Nahyan, who had always treated my father like a brother and me like a daughter. I thought of my father, who used to say, "The wheel of fortune turns, and in its revolution is a better future." General

Musharraf, the man who had sworn to finish me politically, was now going to meet me to discuss the future of Pakistan.

Much to my surprise, the meeting was both long and cordial. We had a one-to-one meeting for several hours. I brought up all the critical political issues, the contentious issues, and General Musharraf's response to all of them was positive. I said that it was absolutely necessary for him to shed his uniform as army chief of staff. I made it clear that there had to be free, fair, and transparent elections that were internationally monitored and that a new, impartial Election Commission must be formed to supervise the elections. I said the elections must be open to the participation of all parties and all party leaders and that procedures must be in place to guarantee not only free voting but accurate counting. I said the ban on twice-elected prime ministers that he had written into the Constitution must be lifted. I said that for true reconciliation, charges brought against parliamentarians from all political parties since he had taken office that had not resulted in convictions, must be dropped. He readily acknowledged that the charges that had been brought against me and my family had been politically motivated and designed to destroy my reputation, and I said he must indicate that not only privately but also publicly.

To my utter surprise he not only agreed in January 2007 to retire as army chief to pave the way for an understanding with the PPP and for national reconciliation but also took me into confidence that it would be done in October 2007, before the presidential election. As the PPP did not recognize the legality of the elections by the existing Parliament, General Musharraf, who had a majority, did not ask for our votes. We decided to leave the legality issue to the courts. It was decided that I would be in Pakistan by December 31, 2007, New Year's Eve, and General Musharraf said he would join me at Bilawal House for New Year's Eve if I invited him. The return issue was big, and it was resolved, too. He promised me to implement what he could constitutionally do through the PPP fair election paper. He agreed to explore a compromise on the problematic issue of the president's power to dismiss Parliament. At the end of the conversation, without my mentioning it, General Musharraf said that for confidence building he would finish the cases that had been lodged against us and gotten nowhere.

Used to fractious, long-winded arguments with opposition parties, I could not believe everything had gone so well. In fact, it had gone too well. Being cautious by nature, I insisted that a schedule be worked out: the first phase should conclude before the presidential elections, the phase for fair elections before the general elections, and the third phase after the elections. I said the PPP would agree to an understanding based on a sequence of events rather than a one-go arrangement.

I was worried: while all the major demands, including retiring as army chief and lifting the ban on twice-elected prime ministers had been agreed to, what were we to do if at the end the general and his team walked away after a rigged election, leaving us high and dry? Therefore I did not want to agree to an understanding on behalf of my party until certain benchmarks were met. General Musharraf and his emissaries kept assuring me that a strategic decision had been made for the moderate forces to work together irrespective of the election results. However, it was the inability to meet benchmarks and to move in the direction of a sequential implementation of what had been agreed to that subsequently led to wrangling and the threat of the agreement coming apart.

General Musharraf and I met again in Abu Dhabi in July 2007. He asked to see me urgently. We were meeting after two important events—first, the removal and reinstatement of Chief Justice Iftikhar, and second, the Red Mosque rebellion, which had left more than a hundred people dead. General Musharraf wanted the PPP's support in reducing the age of retirement for members of the superior judiciary. I could not agree to this, believing it would start a new round of confrontation with the judiciary. Second, he said that he could not finish the cases as he had promised. I told him it was okay if he could not finish the cases. Perhaps he could instead lift the ban on a twice-elected prime minister to show movement? He agreed. Again nothing happened.

In August I called PPP leaders to New York. There we discussed giving General Musharraf a "nonpaper" of what we expected. Makhdoom Amin Fahim gave the "nonpaper" to General Musharraf on August 18. The "nonpaper" said that unless there was movement, by the end of August both sides would be free to go their own ways. General Musharraf and I had a long conversation over the phone that night. He said he would send a team to see me at the end of August.

The August team met me in London at my flat in Queens Gate. They discussed a whole new constitutional package. We increased the political price for the new package. They said they would come back in two days. They didn't. As the deadline approached for calling off talks, I got a call that the deadline would be extended. It was, but there was silence from the Musharraf camp.

The PPP and I met in London in September, and I announced that the date of my return to Pakistan would be given on September 14, 2007, from all the capitals and regions of Pakistan. I wanted the date announced from my homeland. The talks with Musharraf remained erratic. He didn't want us resigning from the assemblies when he sought reelection. There wouldn't be much difference in his winning whether we boycotted or contested, but we used this to press him to retire as army chief. He cited judicial difficulties. It was a harrowing period. After many, many late-night calls, he passed a National Reconciliation Order, rather than lift the ban on a twice-elected prime minister seeking office a third time, which he said he would do later. In exchange for the NRO, we reciprocated by not resigning from the assemblies, although we did not vote for him. We knew the matter still had to be decided by the Supreme Court. We thought Musharraf took the wrong decision to seek reelection from the existing Parliament, that it would only compound the crisis. But he had made his choice.

The NRO was taken up by the courts. They did not stay it but left it undecided for a future date. The ruling party crowed that they had "tricked" the PPP, but we knew that they had only complicated the political situation for General Musharraf and for the country. Following the Supreme Court intervention in admitting the NRO for hearing, none of the persecution against my party or others ended. It upset me to read in the press that I had returned to Pakistan after General Musharraf finished the politically motivated cases against me. None of them was finished. Of all the hearings against my husband and myself that had concluded in the previous eleven and a half years, we had won every one on merit and not on the basis of the NRO.

General Musharraf had agreed to allow free and fair elections under international monitoring. He agreed to an independent Election Commission. He agreed to create a neutral caretaker government. He agreed to allow the newly elected Parliament to revise the rather absurd prohibition against Pakistanis being elected more than twice as prime minister. (This should not be confused with the concept of "term limits" under the presidential system in the West. I was elected for two five-year terms, thus elected to serve for ten years. But the intelligence agencies had twice brought my government down, and I had served for a total of only five years as prime minister. General Musharraf admitted in his autobiography, published in 2006, that the only reason he had rammed through the two-term prohibition was to ensure that I could never be elected prime minister again.) The general also agreed to drop unproven charges against parliamentarians of all major political parties that had not led to convictions in the eight years of his dictatorial rule. If Musharraf had fulfilled his promises, Pakistan could have had an orderly democratic transition, closing the chapter on military rule once and for all.

But as the poet T. S. Eliot once wrote, "Between the idea and the reality falls the shadow." Was General Musharraf merely stalling for time to try to consolidate power? I can't say. But on November 3, 2007, he declared martial law by suspending the Constitution. It put him on a collision course with both the people of the country and the PPP.

All through the years of the Soviet Empire, its Politburo had conducted so-called elections. Calling something an election and actually having an election are obviously very different things. When I was imprisoned in Lahore, under house arrest, surrounded by 4,000 Pakistani militia with bayonets drawn, it was clear that free elections were impossible. Justices, lawyers, political activists, journalists, human rights activists, and students were in prison or under house arrest. The independent media were shut down. A foreign journalist was expelled. The political opposition of democratic parties was banned from campaigning and organizing and denied access to the media.

However, for once domestic pressure and international reaction synchronized. British foreign secretary David Miliband spoke to me, as did U.S. senator Joseph Biden, chairman of the Senate Foreign Relations Committee; U.S. deputy secretary of state John Negroponte; and others. The international community was also calling for the lifting of the emergency, setting a date for elections, General Musharraf's keeping his commitment to retire as army head, releasing of political prisoners, and lifting of the gag on the media.

On November 16, 2007, John Negroponte visited Pakistan. We spoke on the phone. He wondered whether it was possible to put the derailed democratic transition back on track through dialogue.

I shared with Mr. Negroponte the PPP's concern that the regime was using the dialogue process to discredit the opposition without actually making concessions. Though the lifting of the emergency, the retirement of General Musharraf as army head, and setting a date for elections were important, it was also important that an independent Election Commission deal with the nuts and bolts of what made an election fair. I thought that if General Musharraf would respond by lifting the emergency rule (which was, as noted, tantamount to martial law), retiring as army chief, sticking to the announced date of the elections, and taking specific and concrete steps to guarantee the sanctity of the voting and counting process, the political tension in the country could be reduced, enabling a peaceful transition to democracy.

General Musharraf, in response to the domestic and international pressure, announced that elections would go ahead on January 8, 2008. This put to rest the Pakistan Muslim League (Q)'s plans to postpone elections by two years even as the militants advanced from the tribal areas into Swat, the beautiful valley in Frontier Province.

Much to the nation's relief, General Musharraf did retire as army chief, did give a date for lifting the emergency on December 16, and stuck to the January date for elections. However, he did not move on the Election Commission issue.

The PPP continued to get reports of widespread rigging preparations. These included bulk transfers and postings of the judicial officers who have custody of ballot papers. We worried about ghost polling stations as well as the use of government funds in aid of government-sponsored candidates.

To build pressure for the holding of fair elections, I met with former prime minister Nawaz Sharif on December 4, 2007, at my house in Islamabad. I tried to convince him and his allies to participate in the elections, believing it would be more difficult to rig against all of us.

At times I worry whether we as a nation can survive the threat of disintegration. Since the overthrow of my government, the militants have made many inroads into the very structure of governance of Pakistan through their supporters and sympathizers. Pakistan is a tinderbox that could catch fire quickly. It is my home, the home of my children, the home of all the children of Pakistan for whom such enormous sacrifices have been made.

Sixty years after Pakistan's creation, the case study of our nation's record with democracy is a sad chronicle of steps forward and huge steps backward. But this too will change. In the words of the great Pakistani poet-philosopher Iqbal, "Tyranny cannot long endure."

5

Is the Clash of Civilizations Inevitable?

Although many people believe the epic debate over the inevitability of a confrontation between Islam and the West was triggered by Samuel Huntington's essay in the journal *Foreign Affairs* in 1993, the battle of words and ideas on this issue has been raging within and outside the academy for a better part of a century. Actually, this is the critical question before us, for if the conflagration cannot be avoided, the world must prepare for it; but if the confrontation can be reconciled, we must exhaust all remedies to resolve it.

The intellectual debate is represented by two separate and distinct schools. I have chosen to call those who believe in the inevitability of the conflict "clashers," and those who believe the contrary, "reconciliationists." Clearly I am a reconciliationist, and I will outline a program of reconciliation later in this book. But before we talk about a way out, it is important to examine the intellectual frameworks of both sides of the debate.

·

The opening blow in the battle of ideas about the inevitability of the clash of civilizations was struck by Oswald Spengler in his opus *The Decline of the West,* first published in 1918 at the conclusion of World War I. Spengler defines history by civilizations (he actually refers to them as cultures). This categorization had been made before, but it is

Spengler's piece that is seen as the critical intellectual defining element of the debate. Spengler believed that there have been eight "high cultures" in the history of human beings. He believed that the then-current (1918) dominant culture in the world was Western (Spengler named it Faustian), which had begun one thousand years before in Western Europe and had entered a state of decline.

Spengler divided each dominant culture through history into four metaphoric seasonal stages of rise and decline. Spring is the beginning of a culture—the point of creation. Summer is the phase of a culture when civil society is created and critical thought develops. Autumn is the height of a culture's contribution and power. Finally, a culture crumbles in the winter, when new critical thought runs dry and what Spengler calls "irreligiosity" takes over society.

I believe there are two elements of Spengler's work that are important. Spengler made civilizations a term of reference for historical study. This is how others could later argue that civilizations defined history. This element is crucial to Huntington's argument and to the thrust of the entire Huntington school.

Spengler's book became popular in the era immediately following World War I. Then, after World War II, the thesis of his book—that the decline of the West was under way—seemed to be increasingly prescient. The two world wars seemed to reinforce Spengler's conclusions. Actually, Spengler's popularity after each of the world wars is comparable to the enormous revival of interest in the work of Samuel Huntington immediately after the attacks of September 11, 2001, fully eight years after his book *The Clash of Civilizations* was first published.

Although Arnold Toynbee was writing at the same time as Spengler, his multivolume grand review, *A Study of History,* was not completed until 1961. Toynbee's work continues and amplifies the civilization-based approach to history that Spengler put forward. His view is that the downfall of a civilization comes when the "creative minority" simply stops producing good ideas and becomes satisfied to rule in an unjust manner. At that point the "creative minority" becomes merely a "dominant minority" and precipitates the decline of that culture's dominance.

Like Spengler, Toynbee argues that civilizations rise and fall in cycles, naming twenty-three civilizations through history that conform to his theory. Toynbee is important to us because he continues to develop the theory of civilization-based history. His work had a massive popular following (he was on the cover of *Time* magazine in 1947), and thus his theories became similarly popularized, despite the fact that his work was criticized by scholars for being too based in Christian morality rather than on historical fact. Nevertheless, his work helped set the stage for the intellectual developments of the more specific "clashers" who would follow him.

The "clash" academic who shaped the school most directly was Bernard Lewis, a prominent scholar of the Middle East and Islam. Lewis wrote the 1990 article "The Roots of Muslim Rage," in which he actually coined the term "The Clash of Civilizations," which would later gain prominence from Huntington's article by that name.

Lewis believes that Muslim countries have tried Western economic and political models but that the result has been poverty and tyranny. It is quite rational, then, for Lewis to conclude that not only do Muslims reject the models of the West as universal systems, but they will continue to reject them. He believes that Muslims have a major problem with Western ideas of secularism and modernism. Both are seen as having negatively transformed the Muslim world over the past century by destroying its values and social structures.

The United States and the West are seen as responsible for these twin "evils" and thus are legitimate targets for those in the Muslim world who believe in confrontation, not cooperation:

> It should by now be clear that we are facing a mood and a movement far transcending the level of issues and policies and the governments that pursue them. This is no less than a clash of civilizations—the perhaps irrational but surely historic reaction of an ancient rival against our Judeo-Christian heritage, our secular present, and the

worldwide expansion of both. It is crucially important that we on our side should not be provoked into an equally historic but also equally irrational reaction against that rival.

The "clash school" found its strongest advocate and most renowned theorist in a professor who taught at Harvard while I studied there in the early 1970s. Samuel Huntington published "The Clash of Civilizations?" in the journal Foreign Affairs *in the summer of 1993. The essay set off a firestorm of interest, controversy, and debate in the international relations community in the United States and all around the world. "The Clash of Civilizations?" would become a seminal work among the many produced by senior academics after the implosion of the Soviet Union, outlining the parameters and modalities of international relations absent the relatively simple polarities of the Cold War.*

Huntington's work, initially derided by many, has become more and more accepted in the half decade since the September 2001 terrorist attacks on the United States; the March 11, 2004, train attacks in Madrid; the July 7, 2005, subway attacks in London; and the internationalization of terrorism, which had, during the Cold War, been confined to local disputes within nations or regions (Palestine, Kashmir, Chechnya). With the eruption of global international terrorism by Islamic extremists against Western targets, many observers believed that events had validated the thesis of "The Clash of Civilizations?" Indeed, some took to describing the battle between religious terrorist extremists and the United States and Europe as a clash of civilizations. The theory is critical to our current study. Its ominous central point, about the bipolar nature of world conflict becoming a series of multiple clashes between civilizations, is now prominent in policy and academic discussion. It is therefore important to revisit its arguments fifteen years after its initial publication. I want to look at Huntington both within the current context of international terrorism directed against the West and in regard to the internal battle within the Islamic world between moderation and extremism.

But let me be clear from the outset: I disagree with the thesis in "The Clash of Civilizations?" and I fear that this work has actually helped provoke the confrontation it predicts. It is a self-fulfilling prophecy of fear that disregards history and human nature, molding the world to conform to a theory. The assertions about Islam are widely misinformed. The notion that the culture of Islam is antithetical to democratic values not only is unsubstantiated by quranic reference and Islamic clerical interpretation but also plays into Islamic extremists' views that the West is disrespectful and antagonistic to Islam's beliefs and history. The assertion that social, political, and economic interactions between Islam and the West precipitate conflict, as opposed to promoting understanding, is antithetical to what we know about human behavior and international relations. Such a thesis could predispose the West to accepting the "inevitability" of armed conflict with Islamic societies. It shamelessly prescribes methods for the West to hold Islamic nations in check by denying them the tools of modernity and technology. The clash of civilizations theory is not just intellectually provocative: it fuels xenophobia and paranoia both in the West and in the Islamic world. I believe its methodology is flawed, and its conclusions are historically unsupported. Its morbid conclusions are certainly not inevitable.

The central concept of Samuel Huntington's clash of civilizations thesis is that the post–Cold War world has entered a new era of conflict and that this new era will be very different from the past few hundred years of conflict. Conflict will be between civilizations, between fundamental values. In other words, in the language of the title, conflict in the post–Cold War era will be based on civilizations. The world is divided into several specific groupings of civilizations that will replace the identification with nation-states. The bipolar world of the Cold War has been replaced by a new international era that can best be understood as a division of the world into seven or eight groups of civilizations and in which conflict will be defined by a conflict of values among them.

It is notable that the theory argues that civilization affiliation has actually been the basis for conflict for the preponderance of world history. The Cold War era is seen as an aberration in the narrative of history. The thesis posits that international conflicts were predominantly focused in the West during the last few hundred years, when the entire thrust of international relations dealt with interactions within the communities and nations there.

Huntington explains that there were three eras outside traditional civilization-based conflict during this period of Western exception: the era of princes, during which conflict was generally fought between monarchs and other absolute rulers; the era begun by the French Revolution, an era during which nation-states in the West emerged as the primary actors in international conflict; and, lastly, the era of ideologies, characterized by ideological groupings of states acting in concert. During World War II, the Allies and the Axis were not just different regional groupings in the context of war: ideologies such as fascism, liberal democracy, and communism grouped nations together on the battlefield. After World War II ended, the world was once again clearly and absolutely defined by the polarized ideologies of the Cold War.

The clash school argues that, in the current post–Cold War era, the battle of ideologies is over. Replacing ideological demarcations, the new world international stage is defined by conflicts between civilizations. The emergence of civilization as the predominant source of the new era's world conflicts coincides with the emergence of non-Western nations (e.g., Japan, India, and China) as significant players in international politics. In this view, the new era is marked by the dramatic emergence of the power of the non-West.

I think it's important to explore this complex construct. First, we must define the very concept of civilization. To Huntington, civilization is "the highest cultural grouping of people and the broadest level of cultural identity people have short of that which distinguishes humans from other species." Each civilization is defined by certain critical characteristics, including religion, language, customs, and institutions. Civilizations, in this construct, can vary in size from very large (Islamic civilization) to relatively small (Japanese civilization). Although civilizations border one another geographically, these borders are not demarked by sharp, defined lines, but rather often by gradations of influence, by gradual shifts in population and values, and by transition zones.

Huntington identifies seven distinct current civilizations: Western, Confucian, Japanese, Islamic, Hindu, Slavic-Orthodox, and Latin American. He argues that there are subsets and subgroups within each of these defined civilizations: for example, "a resident of Rome may define himself with varying degrees of intensity as a Roman, an Italian, a Catholic, a Christian, a European, a Westerner."

The clash of civilizations theory provides a number of reasons why future world conflicts will occur between civilizations and "along the cultural fault lines separating these civilizations from one another." It argues that "differences among civilizations are not only real; they are basic." A civilization is defined by religion, language, customs, traditions, and institutions. These characteristics do not develop in the short term, nor do they develop in a cultural vacuum; they are the products of centuries of economic, political, and cultural history.

And although the characteristics of any civilization can change and adapt over time, they do not disappear. Significant variance on any of the variables among different civilizations can lead to conflict. If the conflict is serious and prolonged, it can, of course, lead to violence and even war. And although differences over the elements of civilization do not necessarily lead to conflict, the clash theory argues that, historically, "differences among civilizations have generated the most prolonged and the most violent conflicts."

In Huntington's view, "cultural characteristics and differences are less mutable and hence less easily compromised and resolved than political and economic ones." He insists that while ideological differences can be compromised, cultural differences cannot. Capitalists can become Communists, but a Frenchman cannot become a Nigerian, nor can someone from China become Brazilian. Continuing along these lines, he argues that "even more so than ethnicity, religion discriminates sharply and exclusively among people." He warns that "a person can be half-French and half-Arab and simultaneously even a citizen of two countries. It is more difficult to be half-Catholic and half-Muslim." The clash of civilizations sees religion as the ultimate defining variable and thus the ultimate "separator" between people.

The second explanation for the inevitability of conflict in clash theory is the rapidly developing process of globalization. Globalization—the interconnectivity among economies, polities, and societies—is one of the most critical changes in international relations of the modern era. With the advent of faster and more efficient communication and transportation systems, the world is getting smaller, and people, wherever they live and whatever they do, are growing less isolated. While most argue that interactions can break down barriers and bring people closer, the Huntington school unpersuasively argues the opposite, that these interactions "intensify civilization consciousness and awareness of differences between civilizations and commonalities within civilizations." He believes that globalization promotes xenophobia and cultural contempt rather than pluralism and tolerance.

Huntington points to economic factors as the third factor in the inevitability of the clash. Economic modernization is affecting the world as never before. Partnered with the social change that it sets in motion, economic modernization separates people from their "long-standing local identities." Traditional life is altered, and what historically defined a group of people no longer exists. Thus, the theory argues, people increasingly turn to fundamentalist religion, often to fill the resulting gap. People want continuity and stability. They want to be grounded. When the traditional elements that defined their daily lives are disrupted, they naturally attempt to fill the social and psychological vacuum and instability in their lives.

One of the most significant ways in which this vacuum can be filled is with the assertion of strong religious values, beliefs, and practices. Religion thus fills the gap left by the elimination of traditional society. This new, stronger adherence to religion deepens perceived differences between civilizations and strengthens perceived similarities between members of the same civilization.

A fourth factor in the clash of civilizations theory concerns the "dual role of the West." Civilization consciousness is becoming more important because the West is at the peak of its power, but at the same time, many non-Western civilizations are experiencing what is described as a "return to the roots phenomenon," a rejection of Western values and a turning inward to embrace indigenous values. It is a rejection of the values and symbols of Western culture—the music, the clothes, the art, and the structures of governance. Non-Western civilizations are decreasingly looking to the West as a model of development and modernization while they feel they can develop in a distinctly non-Western way. Clashers believe this rejection of Western values, at the same time as the West is at the peak of its economic, political, and military power, causes tension and ultimately confrontation. The confrontation will be between the West and those who increasingly have the "desire, the will and the resources to shape the world in non-Western ways."

Huntington describes some of the areas of the world where civilization-based conflict is occurring and predicts where such clashes are likely to occur in the future. In the age of civilization consciousness, these "fault lines" are the areas where one civilization's border meets another. Though Huntington uses examples of conflict in the former Yugoslavia, China, and Japan, he focuses on conflicts related to Islamic civilization because "Islam has bloody borders." He states that conflict between Western civilization and Islamic civilization has been ongoing for more than 1,300 years and that in fact "this centuries-old military interaction between the West and Islam is unlikely to decline. It could become more virulent."

The clashers argue that—unlike for the West—democracy is dangerous for the Muslim world and ultimately increases the likelihood of the clash of civilizations because the "principal beneficiaries of these [democratic] openings have been Islamist movements." "In the Arab world, in short, Western democratic practices strengthen anti-Western political forces." (Note that democracy is synonymous with "Western democracy"—the theory dismisses the universality of the values of freedom, democracy, and liberty.) Huntington goes on to cite examples of conflict along Islamic civilization's southern border with "animist/pagan/Christian Africa" in places such as Sudan, Chad, and Nigeria. On Islam's northern border, he cites conflicts with people of the Eastern Orthodox Church, including Serbs and Albanians; Bulgarians and the Turkish minority; Ossetians and Ingush; and Russians and Muslims in Central Asia. In the East, Islamic civilization clashes with Hindu civilization. In fact, according to the clash of civilizations theory, "wherever one looks along the perimeter of Islam, Muslims have problems living peaceably with their neighbors."

Having claimed—I would argue incorrectly—that Islamic civilization conflicts with other civilizations more often than other civilizations fight one another, he suggests several explanations for the increase in conflict between Islam and the West. He argues that Muslim population growth has put pressures on Muslim societies (unemployment and inadequate infrastructure) and on the Western societies to which they emigrate. He explains that a recent "Islamic Resurgence"—an upswing in Islamic religiosity around the world—has given Muslims renewed confidence in their own values and the superiority of their culture over that of the West. And he argues that "the West's simultaneous efforts to universalize its values and institutions, to maintain its military and economic superiority, and to intervene in conflicts in the Muslim world generate intense resentment among Muslims."

Finally, he argues that increased contact between the two civilizations helps develop a new sense of identity within each and thus highlights the differences between them. "The Clash of Civilizations?" concludes by observing that "so as long as Islam remains Islam (which it will) and the West remains the West (which is more dubious), this fundamental conflict between two great civilizations and ways of life will continue to define their relations in the future even as it has defined them for the past fourteen centuries."

The essay identifies what it calls the "Kin Country Syndrome." It explains that countries that are within the same civilization ("kin countries") will turn to one another for support during conflicts. Huntington points to the first Persian Gulf War as an example of the syndrome. During this conflict, the Iraqi dictator Saddam Hussein attempted to frame the war as a battle of civilizations, Islam versus the West. It was a shaky assertion at best, an absurd one at worst, since Saddam provoked the conflict by invading Kuwait, an independent, sovereign fellow Muslim country. Huntington asserts that the political coalition of Islamic states that supported the Americans soon fell apart after the war due to the Kin Country Syndrome. He even claims that during the war most Muslim elites supported Saddam Hussein, even if that support was manifested only through "private cheering." And of course there were public demonstrations of support for Saddam in Jordan and Palestine.

"The Clash of Civilizations?" argues that this and other examples show the increasing importance of the idea of "civilization" in world conflict. It does not, however, argue that other types of conflict will not take place. Conflict may occur in the future that is not intercivilizational, but "such conflicts . . . are likely to be less intense and less likely to expand than conflicts between civilizations."

Huntington argues that many of the future civilizational clashes will occur in a West-versus-non-West framework. Currently the West dominates the world through its economic, political, and military power. It uses international institutions to promote its own interests, at the exclusion of others. With the West at the peak of its power, there will be two types of disputes that will fuel conflict in a civilization-based context: battles based on the struggle for power in military, institutional, and economic interests and conflict based on "differences in culture . . . basic values and beliefs."

The cultures of Western and non-Western civilizations will continue to clash, especially in areas of "individualism, liberalism, constitutionalism, human rights, equality, liberty, the rule of law, democracy, free markets, and the separation of church and state." Specifically, he argues that the values of secularism are particularly offensive to Islamic culture: "In Muslim eyes Western secularism, irreligiosity, and hence immorality are worse evils than the Western Christianity that produced them." In many cases this produces and will continue to produce an indigenous backlash as the West's cultural imperialism collides with non-Western native cultures. Huntington focuses on democracy and concludes that "it is striking the relative slowness with which Muslim countries, particularly Arab countries, have moved toward democracy. Their cultural heritage and their ideologies may be in part responsible."

Huntington believes that intercivilizational conflict is the type of conflict that is most likely to lead to global wars—and that "a central focus of conflict for the immediate future will be between the West and several Islamic-Confucian states."

He prescribes a number of short- and long-term policy prescriptions for the West, some especially relevant to our discussion. He argues that in the short term the West should prevent localized intercivilizational conflicts from escalating; limit the military expansion of Confucian and Islamic states; slow the reduction of Western military capabilities; maintain military superiority over Confucian-Islamic states; strengthen international institutions that promote Western values; strengthen Western interests and promote membership of non-Western states into those institutions; support groups within other civilizations that are sympathetic to Western values; and "exploit differences and conflicts among Confucian and Islamic states." "The West will increasingly have to accommodate these non-Western modern civilizations whose power approaches that of the West but whose values and interests differ significantly from those of the West. This will require the West to maintain the economic and military power necessary to protect its interests in relation to these civilizations."

A number of prominent thinkers agree with Huntington that Western civilization and Islamic civilization are in a pitched battle and that the clash of civilizations that Lewis first named and Huntington predicted either is imminent or has already begun. This list includes individuals on both sides of the falsely constructed divide— pundits, clerics, and (frankly) paranoids. Their modus operandi is lobbing intellectual and rhetorical bombs into an already incendiary debate, preparing for a war that they believe (and maybe hope) is inevitable.

Let's start with an American pundit who, some might argue, occasionally crosses the line into outright xenophobia, Patrick Buchanan. This well-known conservative politician often seems to argue that a clash of civilizations is coming and that it is imperative that the West win the battle. In Buchanan's view, the only way the West can "beat" Islam is for the West to regain religious faith. But at the present time, Buchanan actually thinks the West would lose such a battle with Islam. "To defeat a faith, you need a faith. What is ours? Individualism, democracy, pluralism, la dolce vita? Can they overcome a fighting faith, 16 centuries old, and rising again?"

Robert Spencer is the author of the well-known Web site Jihad Watch. He uses the Internet to spread misinformation and hatred of Islam, while claiming that he is merely putting forward the truth. But as in much extremist advocacy, he presents a skewed, one-sided, and inflammatory story that only helps to sow the seeds of civilizational conflict. For example, he takes apparently violent verses of the Quran out of context and then does not provide any peaceful verses as a balance.

Unlike many of the more mainstream authors presented, Spencer does not understand the true Muslim faith or differentiate between moderate Muslims and violent Islamists, and so lumps them all in one boat:

> *Islam is a totalitarian ideology that aims to control the religious, social and political life of mankind in all its aspects, the life of its followers*

without qualification, and the life of those who follow the so-called tolerated religions to a degree that prevents their activities from getting in the way of Islam in any way. And I mean Islam; I do not accept some spurious distinction between Islam and "Islamic Fundamentalism" or "Islamic terrorism." The terrorists who planted the bombs in Madrid, and those responsible for the death of more than 2000 people on September 11, 2001 in New York and the Ayatollahs of Iran were and are all acting canonically; their actions reflect the teachings of Islam, whether found in the Koran, in the acts and sayings of the Prophet, or Islamic law based on them.

The intellectual radicals in the clashers' divide are not all Western. Maulana Maudoodi, the founder of the Jamaat-i-Islami Maudoodi, believed that all nations that stood in the way of Islamic Sharia rule should be eliminated through violent jihad, including the West. Thus he saw the West and Islam in perpetual and irresolvable conflict. His view of the West is as one-sided and distorted as the Western clashers' view of Islam.

It is a purely materialistic civilization. Its whole system is devoid of the concepts of compassion, fear of God, straightforwardness, truthfulness, urge for the right, morality, honesty, trustworthiness, virtue, modesty, piety and chastity which form the foundations of Islamic civilization. Its ideology is diametrically opposed to the Islamic concept. It leads mankind in a direction contra Islam. Western civilization strikes at the roots of that concept of ethics and culture which is the base of Islamic civilization. It builds its individual and collective character on a pattern that cannot afford and adjust with the structure of Islamic civilization. In other words Islam and Western civilization are like two boats sailing in totally opposite directions. Any attempt to sail in both the boats at a time shall split the adventurer into two pieces.

Sayyid Qutb provided the intellectual foundation of the Muslim Brotherhood in Egypt in the 1960s. The Brotherhood is considered by many to be the ideological father of Islamic Jihad, Al Qaeda, and many other modern terrorist organizations. A powerful force in grass-roots, populist Egyptian politics, Qutb was seen as a threat to the Egyptian government who had to be eliminated, and he was hanged for his views. "The leadership of mankind by Western man," he wrote, "is now on the decline, not because Western culture has become poor materially or because its economic and military power has become weak. The period of the Western system has come to an end primarily because it is deprived of those life-giving values which enabled it to be the leader of mankind."

This quote seems as if it could be taken straight from Toynbee. The West is in decline because its leaders have run out of creative/valuable ideas and has gone from being the "creative minority," which ruled based on merit, to the "dominant minority," which seeks to extend its rule. Qutb believed that "it is necessary for the new leadership to preserve and develop the material fruits of the creative genius of Europe, and also to provide mankind with such high ideals and values as have so far remained undiscovered by mankind, and which will also acquaint humanity with a way of life which is harmonious with human nature, which is positive and constructive, and which is practicable." Qutb was the quintessential advocate of the civilizational argument but came at it from the opposite ideological pole as Huntington. To Qutb the West is a dying civilization that needs to be replaced.

Khurshid Ahmed was a prominent Pakistani professor and scholar who viewed the clash of civilizations as an inevitable consequence of the West's determination to impose its values, culture, and forms of government on the Muslim world: "If in the Muslim mind and the Muslim viewpoint, Western powers remain associated with efforts to impose the Western model on Muslim society, keeping Muslims tied to the system of Western domination at national and international levels and thus destabilizing Muslim culture and society directly or indirectly, then, of course, the tension will increase. Differences are bound to multiply."

The Hizb ut Tahrir, an Islamist organization based in the United

Kingdom, believes in the establishment of a classic caliphate, a Muslim political and religious empire stretching from Southern Europe through the Middle East into South, Central, and East Asia:

> *It is the height of cultural hubris that only the west's chosen political system, secular democracy can build an effective society. This view also ignores completely the fact that the Islamic political system, the Caliphate also delivers a system that can bring representation, accountability, a rule of law and which redistributes wealth. Yet despite centuries of evidence of Islam's effective rule, and the abundant realities of secular democracy's failings, many in the west still prefer to believe the emperor's naked body remains fully clothed. What Georgia, Lebanon, Ukraine, Pakistan, Russia and numerous other cases show us, is that secular democracy is simply not up to the test.*

Ayaan Hirsi Ali is an extremely controversial Somalian-born former Dutch parliamentarian who seems to endorse the clash theory because she has so completely come to reject Islam as a positive force: "But I could no longer avoid seeing the totalitarian, the pure moral framework that is Islam. It regulates every detail of life and subjugates free will. True Islam, as a rigid belief system and a moral framework, leads to cruelty. The inhuman act of those nineteen hijackers was the logical outcome of this detailed system for regulating human behavior. Their world is divided between 'Us' and 'Them'—if you don't accept Islam you should perish." Her attacks on Islam and even on the Prophet himself have I believe put her outside the rational, useful debate over competing cultures. Now a resident scholar at the American Enterprise Institute, a conservative think tank in Washington, she seems to have been adopted by neocon clashers as a "poster child" for the inevitability of the clash of civilizations.

•

Clearly the clash of civilizations thesis is one of the seminal theories in international relations in the post–Cold War era. *Foreign Affairs,* the original publisher of Huntington's paper, reported that the work was the most commented-on article of any it had published since the 1940s. Many prominent authors have disagreed with all or part of Huntington's thesis and have written various critiques of his work. Now that we have discussed the theory at length, let us try to systematically break it down (and break it up). I will present my own views on its distinct intellectual problems and some related prescriptions, following a review of some excellent academic critiques.

Stephen Walt is the author of one of the best-written and most comprehensive refutations of the clash of civilizations thesis. Given the simplicity of Huntington's thesis and the enormous subject area that it aims to cover, one approach is to challenge significant portions of the argument and simultaneously chip away at other less central pieces of the theory. In most critiques, the attacks fall short of a successful frontal assault on the main theses of the clash of civilizations and fail to offer convincing alternative postulations of the emerging factors that will shape international relations in the post–Cold War era. Walt, however, manages both the negative and positive—criticizing its view and offering a credible alternative—with grace and coherence.

Walt's critique of the clash of civilizations argues that the theory does not properly explain why loyalties in the post–Cold War era are actually shifting from the national level to the civilizational level. Walt believes that the legitimacy of the thesis is not demonstrated with academic rigor. Second, even if one assumes that Huntington is correct that the clash of civilizations will be a result of the shift of loyalties from nations to civilizations in the post–Cold War era, he does not demonstrate why this shift in loyalties will necessarily lead to violence.

Walt argues that states have always acted in their perceived self-interests: "For the past 2,000 years or so, assorted empires, city-states, tribes and nation-states have repeatedly ignored cultural affinities in order to pursue particular selfish interests." Walt reminds us that states have been acting in their own self-interest for centuries and conflicting with other states outside their own civilization. The thesis fails to satisfactorily answer the question why identity in the international arena switches from the national level to the civilizational level. Walt's principal problem with the clash of civilizations theory is its inability to explain why conflict is more likely to happen between civilizations than within them. Walt demonstrates that cultural differences between civilizations do not necessarily lead to conflict, "just as cultural similarities do not guarantee harmony." Even if two civilizations are different in significant aspects, conflict is not the only possible outcome of their interaction. In fact, one could argue that cultural diversity could make the world more stable. Certainly the guiding principle of "public diplomacy" is that familiarity with other cultures through exchanges of students, scholars, government officials, scientists, doctors, and engineers leads to respect for cultural diversity and understanding. It is the practical consequence of diversity that is central to educational and cultural exchange between nations. My personal experience suggests to me that educational and cultural exchange leads not to conflict but to the opposite. My years of studying and lecturing abroad have not only sensitized me to other cultures and societies but have sensitized those I have interacted with on all levels to my culture, nation, and religion. My life experience thus demonstrates that Huntington's theory is incorrect, and I will amplify this later.

Walt's final major criticism of the idea of the clash of civilizations is of the academic methodology used in the paper. He provides a litany of holes that he believes undermine the theory. One example that he uses is that of the Persian Gulf War. Huntington concedes that during this conflict a coalition of Western and Islamic states was formed to repulse an Islamic state from occupying another Islamic state. Both this conflict (the intracivilizational violence) and the broad-based coalition (intercivilizational cooperation) undermine the Huntington theory. But defenders of the clash of civilizations theory counter by making the claim that while some Islamic states participated in the coalition, elites in these Muslim countries privately cheered the Iraqi leader, Saddam Hussein, and his efforts, especially as he resisted the West's military intimidation both in the air and on the ground.

For Walt, this example reinforces his argument that states' individual interests trump their civilizational identity. The fact that the states acted in self-interest—while many elites felt a civilizational pull—underscores the reality that state interests mattered more than loosely felt and politically impotent loyalties to a particular civilizational entity. Walt makes the point that "in the Gulf war, in short, civilizational identities were irrelevant."

Walt concludes that the present condition of international relations is not a clash of civilizations but a continuation of the pattern of nation-state conflict that has characterized international relations in the modern era. He lists conflicts involving the Abkhaz, Armenians, Azeris, Chechens, Croats, Eritreans, Georgians, Kurds, Ossetians, Quebecois, Serbs, and Slovaks as evidence of this trend of national and state interest being the primary driving force in international conflicts. The reason that nation-state self-interest remains the primary source of conflict is that states are the primary actors in the world; they "can mobilize their citizens, collect taxes, issue threats, reward friends, and wage war; in other words, states can act. . . . [T]his neglect of nationalism is the Achilles' heel of the civilizational paradigm." And if the reader is quick to jump to the so-called war on terror as the emergence of intercivilizational conflict to substantiate Huntington's thesis, I would assert that the true war on terror is not between Islam and the West but rather between moderate and extremist forces within the Islamic world. The war on terror reflects intracivilizational conflict, which confounds and contradicts the thesis of emerging intercivilizational conflict as the central reality of the post–Cold War world.

Walt concludes his paper with a warning. The clash of civilizations, if accepted, can lead to future conflicts because "if we treat all states that are part of some other 'civilization' as intrinsically hostile, we are likely to create enemies that might otherwise be neutral or friendly." Thus Walt warns us that "in this sense, 'The Clash of Civilizations' offers a dangerous, self-fulfilling prophecy: The more we believe it and make it the basis for action, the more likely it is to come true." Thus there will be a clash of civilizations only if we allow it by believing that it is inevitable and unavoidable.

Professor Fouad Ajami, of the School for Advanced and International Studies at the Johns Hopkins University, responded to the publication of the "The Clash of Civilizations?" with a critique in the very next issue of *Foreign Affairs*. The central thesis of Ajami's response, concurring with Walt, is that "The Clash of Civilizations?" dismisses the role of the state in international conflict, a role that has been the central driving force in international relations for hundreds of years. While Huntington's argument might be convincing because of its effective and emotional appeal, Ajami believes that states will not act according to civilizational identity at any point in the near future. They will continue to behave consistently with their perceived self-interest. Ajami seems almost surprised that Huntington would even offer the thesis. "From one of the most influential and brilliant students of the state and its core interest there now comes an essay that misses the slyness of states, the unsentimental and cold-blooded nature of so much of what they do as they pick their way through chaos."

Ajami thinks that "The Clash of Civilizations?" was influenced by certain indicators that should have been viewed more skeptically. States, he argues, act in their self-interest, but the leaders of these states will often argue otherwise. He offers the example of the Persian Gulf War, which he believes is not a good example of civilization-based conflict. Ajami seems mystified at Huntington's inability to recognize or accept that the perceived self- (or selfish) interest of states motivated the war.

Instead, "Huntington buys Saddam Hussein's interpretation of the Gulf War." Saddam had declared the war to be a war of Islam versus the West. He emphasized "kinship" to frame the war as a civilizational war to get Islamic states to fight with Iraq. (The Scud attacks on Israel were clearly meant to broaden the perception of the war away from his aggression against Kuwait to an attack on a conveniently common enemy.) Saddam certainly didn't know it, but he was in fact testing Huntington's "kin countries" theory. The "kinship" theory failed the test because states might use the rhetoric of higher terms of "kinship" but in fact act out of state self-interest. In the end, this example "laid bare the interests of states, the lengths to which they will go to restore a tolerable balance of power in a place that matters." Islamic states did not heed Hussein's civilization-based call but stood their ground and fought for their state-based interests.

The Balkan example is also skewered by Ajami. He believes that the violence in the Balkans was not a clash of civilizations along a civilizational fault line, as argued, but rather that the various factions fighting in the former Yugoslavia were engaged in conflict only to increase their local power. What made the leaders such as Slobodan Milošević and Franjo Tudjman brilliant, according to Ajami, was that they were able to frame "their bids for power into grand civilizational undertakings—the ramparts of the Enlightenment defended against Islam or, in Tudjman's case, against the heirs of the Slavic-Orthodox faith." Once again, Ajami argues that the clash of civilizations theory miscalculates the true nature of a conflict as a result of the public manipulation of facts. The leaders of factions were attempting to justify their aggression and brutality by citing broader historical and cultural explanations that camouflaged the realpolitik of narrow self-interest.

Ajami believes that while the paper's thesis is an interesting academic exercise, state interest remains the primary force that shapes the behavior of nation-states. He concludes, "Civilizations and civilizational fidelities remain. . . . But let us be clear: civilizations do not control states, states control civilizations. States avert their gaze from blood ties when they need to; they see brotherhood and faith and kin when it is in their interest to do so."

The late *Wall Street Journal* editor Robert L. Bartley responded to "The Clash of Civilizations?" in the same issue of *Foreign Affairs* as Ajami. Bartley's main argument is that economic development, the spread of liberal democracy, and the expansion of communications will bring the world closer rather than farther apart. His conclusion is directly opposed to Huntington's, which insists that increased communication will lead to the clash of civilizations resulting from heightened awareness of the differences between them.

Bartley believes that the forces of globalization, which the clash of civilizations theory asserts will drive people apart, will instead bring them together. The clash of civilizations theory sees economic development very differently from Bartley. It posits that economic modernization will lead to the diminution of traditional life. It argues that this loss will lead to a rise in fundamentalism and thus increase conflict between civilizations based on religious competition. Thus modernization correlates with fundamentalism.

I strongly believe, as do many scholars, that modernization will have exactly the opposite effect. I believe that modernization and extremism are contradictory and mutually exclusive and that modernization is related to political and religious moderation and not to fundamentalism. The unsubstantiated assertion that modernization and technology will breed dissonance and conflict and ultimately religious fanaticism is one of the strongest reasons why I believe that the clash of civilizations is incorrect and that it encourages the worst elements of human and state nature.

I agree with those who see economic development as leading to and consonant with democratic development. This "development creates a middle class that wants a say in its own future, that cares about the progress and freedom of its sons and daughters." Thus Bartley believes that if economic development continues, so will democratic development. The development of democracy will help end conflict. Bartley points out that empirical evidence shows the undeniable historical pattern that democracies almost never go to war with one another.

Bartley thus has a positive, optimistic view of the future, believing that "the dominant flow of historical forces in the 21st century could well be this: economic development leads to demands for democracy and individual (or familial) autonomy; instant worldwide communications reduces the power of oppressive governments; the spread of democratic states diminishes the potential for conflict."

This premise provides a rationale for conducting trade with nations that might be politically repressive while being economically pluralistic—the notion that economic development will, over time, open up societies politically. This certainly is the view of many in the West with respect to the potential for political reform within China. The Chinese people are now enjoying previously unheard-of incentives to produce and innovate. A market-based economy seems to be taking hold, even if the government does not characterize it that way. Young Chinese have much broader educational opportunities than ever before, and all people, especially in the urban centers, have extraordinary new economic options. Wealth is being made and accumulated. Families are buying housing, automobiles, and consumer goods. Their use of the newest forms of communication technology now links the people of China to the economic globalization of the planet. In time the economic reform that is sweeping the nation will inevitably result in political reform.

Two outstanding studies use empirical data to determine whether the clash of civilizations thesis has historical support and if it successfully describes the post–Cold War era (1988–1992). In the Russet study, by Bruce Russet, John O'Neal, and Michaelene Cox, the authors conducted an empirical review of conflict, using complex statistical analysis to objectively test the historical validity of Huntington's thesis. Their universe includes conflicts from 1950 to 1992. They test to see if intercivilizational dyads (pairs) of states are more likely to enter into "militarized interstate disputes" than are intracivilizational dyads. They compare the Cold War and post–Cold War periods to see if there was any increase in intercivilizational conflict that would help to prove the validity and accuracy of the "clash of civilizations."

The Russet study finds that the most significant variable that predicts whether two states will enter into conflict is the level of democracy. Based on an analysis of data spanning more than four decades, pairs of democracies were found to be the most peaceful type of dyad. In addition, the study found that economic interdependence "also inhibits conflict modestly." Thus the study disproves two of the most significant elements of the Huntington thesis.

The Russet study then looks at conflict overall to see if more conflicts were intercivilizational or intracivilizational. For intercivilizational conflict, with all the variables accounted for, "there is little evidence that civilizations clash." This is an extraordinary finding that invalidates the major thesis in the clash of civilizations. Four of the civilizations included in the study showed greater incidences of conflict within each civilization than with others. This indicates that civilizations are not prone to clashing but in fact experience mostly internal fighting.

For the time period studied, the information "not only fails to support Huntington's thesis but leans in the opposite direction." Applied to Islam, it substantiates my position that the most significant clash of our modern era is not between Islam and the West but rather is an internal fight within Muslim states, between the forces of moderation and modernity and the competing forces of extremism and fanaticism.

After reviewing the data set as a whole, the Russet study examined whether there has been an increase in the incidence of intercivilizational conflict in the post–Cold War era. As the study looked at data only from the beginning of the post–Cold War era, an increasing trend of intercivilizational conflict might indicate that Huntington's thesis would be proved correct as a trend going into the future.

Once again, however, the data prove Huntington wrong. Using a number of indicators, the Russet study tested to see first if intercivilizational conflict increased as the Cold War decreased in intensity (the intensity of the Cold War was measured by U.S. military expenditures as a percentage of total GDP). As the Cold War's intensity diminished, "conflicts among split dyads were less common . . . the

Cold War seems to have fanned cultural differences, not suppressed them." This conclusion once again contradicts the thesis that the Cold War had a damping effect on the clash of civilizations. The data prove, however, that as the Cold War lowered in intensity, intercivilizational conflict did not increase.

The Russet study also tested to see if the occurrence of intercivilizational conflict increased in the post–Cold War era. The study found just the opposite. The data show that the incidence of intercivilizational conflict actually decreased in the post–Cold War era. Since 1988, a clash of civilizations has not occurred, and intercivilizational conflict has actually decreased as a percentage of all conflict.

The authors give us a glimpse into a possible future far different from the clash school's vision:

Civilizations do not define the fault lines along which international conflict occurs. More relevant are the common bonds of democracy and economic interdependence that unite many states, and separate them from others. The realist influences are important for states that do not share liberal ties. For them, realpolitik still determines the incidence of conflict. Consequently, policy-makers should focus on what they can do: peacefully extending democracy and economic interdependence to parts of the world still excluded.

Much like the authors in the Russet study, Errol Henderson and Richard Tucker conducted an empirical study to see if historical data on actual world conflict prove the clash of civilizations theory correct. As with the Russet study, the Henderson study uses historical data and complex statistical analysis to test its thesis. Henderson looks at the Cold War and post–Cold War eras to see if he can find an increase in intercivilizational conflict in the post–Cold War era as predicted in "The Clash of Civilizations?"

The findings of the Henderson study are consistent with those of the Russet study. In the post–Cold War era, Henderson found, "states of different civilizations are actually less likely than states of the same civilization to fight one another." During the Cold War period, the data are not as categorical; they do not statistically disprove Huntington. The data show that there is not a statistically significant relationship between "mixed civilizations and the probability of war."

Like many of those who critique the clash of civilizations theory, Henderson believes that it "ignores the persistent role of nationalism in world politics"—while the work acknowledges the primary role the state plays in world politics, it ignores the state and the nation as the primary form of identity.

The studies that we have reviewed demonstrate that democratic forms of government serve to inhibit aggressive militarism, especially against other democracies. This is not a normative view of what "should be." It is an objective view of what actually is. According to the data, the greater the number of democratic countries that exist in the world, the less conflict there will be. Thus it is the responsibility of all those who have the power to help bring democracy to those who are without it, to do so. This is not only an ideal, a moral objective, but it is in the self-interest of democratic states to promote democracy around the world and perhaps most of all in areas that have a greater potential for intercivilizational conflict. Providing incentives for democratic transitions in authoritarian states is not only right, it is in the self-interest of the community of nations.

·

But I want to look beyond what Huntington and his school say about clashes between civilizations to look specifically at what the thesis says about Islam. It theorizes that in the post–Cold War world a clash of civilizations will generally occur. It goes on, as we have seen, to say that Islamic civilization is particularly prone to civilizational conflict and, in particular, prone to conflict with Western civilization. Since this analysis and prediction have taken such a critical place in attempting to explain the current confrontation between Muslim extremists and the West, it is important to spend time concentrating on this section of the work. The clash of civilizations theory asserts that

Islam is particularly prone to armed conflict. I will present critiques from academia that contradict this and then go on to comment on it from my own life experiences, including the current political situation in my homeland of Pakistan.

Nothing debunks ideologically driven myths like simple facts, hard evidence. In recent years, there have been two important empirical studies undertaken on the accuracy of the clash of civilizations theory's assertions regarding Islamic civilization. The first important one was conducted by Jonathan Fox in 2001. Looking at the recent history of ethnic conflict in the world, the Fox study examines this set of conflict from "three perspectives: globally, from the perspective of the Islamic civilization, and from the perspective of the Western civilization." The data set is from ethnic conflicts within states, that is, along the civilizational fault lines defined by "The Clash of Civilizations?" He divides his data into two periods: the Cold War (1945–1989) and the post–Cold War era (1990–1998).

Overall, the results from this study do not support the claim that in the post–Cold War era, world conflict will be based on civilizational identity and a major focus of that conflict will be between Western and Islamic civilizations.

First, looking at the picture globally, Fox compared the amount of civilization-based conflict in the world during the Cold War era to that of the post–Cold War era. He found that "overall there is little change in the level of civilizational conflict between the Cold War and the post–Cold War eras. During both periods, about 38 percent of ethnic conflicts were civilizational." So from a global perspective, civilization-based conflict does not make up the majority of ethnic conflict. Additionally, the ratio of civilization-based conflict to the total amount of ethnic conflict in the world did not increase from the Cold War era to the post–Cold War era. Further, Fox looked at the ratio of civilizational conflict that involved both Western civilization and Islamic civilization. The results indicate that of all civilizational conflict in the world, only 14.6 percent was between the West and Islam during the Cold War; and after the Cold War, it was 18.3

percent. That is not a large difference. Once again, if "The Clash of Civilizations?" was correct in its thesis, a sharp rise in civilization-based conflict between the West and Islam would be expected. But the data indicate otherwise.

Second, Fox looked at ethnic conflict from an Islamic perspective. During the Cold War, 63.5 percent of the conflicts that Islamic groups were involved in were classified as civilizational. In the post–Cold War period this number actually decreased to 62.4 percent. According to the thesis in "The Clash of Civilizations?" the ratio of civilization-based ethnic conflicts to ethnic conflicts should have increased in the post–Cold War period. From an Islamic perspective, the data do not validate this prediction.

Third and finally, the Fox study evaluated the data from a Western perspective. In conflicts that involved the West, during the Cold War era, 63.5 percent were civilizational. After the Cold War, the data show a drop, not an increase, in the percentage of Western conflicts that were civilizational: to 59.3 percent. Counter to the prediction in the clash of civilizations theory, from a Western perspective civilizational conflict did not increase in the post–Cold War era.

Overall, during the Cold War 83.1 percent of civilization-based ethnic conflict involved either Western civilization, Islamic civilization, or both. After the Cold War this number dropped to 80.1 percent. Despite Huntington's predictions, the distribution of ethnic conflicts between and within civilizations did not change substantially with the end of the Cold War. Fox admits that Islamic civilization is involved in a large number of conflicts, but this is not new to the post–Cold War era. From a global perspective, conflicts between Western groups and Islamic groups "are a minority of the ethnic conflicts in which Islamic ethnic groups are involved." That two democracies do not often come into conflict with each other is a crucial point for the Islamic world. Obviously, some people argue that democracy and Islam cannot coexist. That is to say that, as Huntington does in his book *The Clash of Civilizations and the Remaking of World Order,*

"This failure [of democracy in the Islamic world] has its source at least in part in the inhospitable nature of Islamic culture and society to Western liberal concepts." As evidenced in the earliest discussion of the theological debate within Islam between democracy and dictatorship and moderation and extremism, I reject the claim that Islam is "inhospitable" to democracy as well as the notion that liberalism is inherently and uniquely "Western." In fact, enough progress can be shown empirically to demonstrate exactly the opposite.

A strong argument can be made that the future of the twenty-first century will be shaped by the peaceful growth of democracy and moderation around the world, but most especially in Islamic nations, countering the thesis presented in "The Clash of Civilizations?" Specifically describing Islamic states is fundamental to the course I have committed myself to and the course of my nation in the months and years ahead. Democracies do not go to war with democracies. Democracies do not become state sponsors of terrorism. I find no evidence to the contrary.

There is a self-interested rationale in the validity of these statements, even disregarding moral issues. The West should be promoting and nurturing democracies in the Islamic world as a matter of national security policy. There will not be a clash of civilizations between Islam and the West if democracy is institutionalized in the Islamic world. Democracy in Muslim-majority states would isolate and marginalize the extremists and fanatics. Not only can democracy coexist with Islam, but it can flourish in Islamic states. That is the cause to which I have devoted my life.

A significant body of evidence suggests that democracy is a universal goal to which both those within the West and within Islam aspire. A 2002 study by the well-known scholars Pippa Norris and Ronald Inglehart compared attitudes toward democratic values. Their study is a "comparative analysis of the beliefs and values of Islamic and non-Islamic publics in 75 societies around the globe." They looked at how Islamic and non-Islamic publics viewed different democratic ideas—

regardless of the system of government under which they lived—in an attempt to identify the level and intensity of support for democratic values in the Muslim world. They found that democracy is a value that is universally yearned for. The support for democracy within the Muslim world is every bit as strong as support for democracy in the West. From my experience on the ground, it is certainly true for the people of Pakistan.

Norris and Inglehart found that "in recent years high support for democratic ideals is almost universally found in most nations around the globe." Democratic values are so universally held that "contrary to Huntington's thesis, compared with Western societies, support for democracy was marginally slightly stronger (not weaker) among those living in Islamic societies." This was indicated by the support for three indicators that Norris and Inglehart tested: the way democracy works in practice, support for the democratic ideal, and disapproval of the idea of strong government leaders.

According to Norris and Inglehart, what splits the two civilizations (Western and Islamic) apart are social issues, not political ones. Western nations, unlike the Islamic world, show a strong preference for the equality of women and the tolerance of homosexuals. Islamic nations demonstrate more traditional social attitudes with respect to these issues. While Islamic and Western societies are different from one another, the differences are less about political attitudes toward elections, governing, and parliamentary democracies than about social and cultural matters. Significantly, Western and Islamic societies both show equal zeal for democratic ideals and democratic governance.

Norris and Inglehart's empirical findings directly contradict one of Huntington's most critical theses within the clash of civilizations theory: that social and cultural differences between Western and Islamic societies create sharp differences between the two societies in political attitudes.

I share the concern of many scholars that the greatest danger of the clash of civilizations thesis is that the theory can become a self-fulfilling prophecy. Bruce Russet is one author who shares this worry:

Similar to other big ideas, it has the potential to become not just an analytical interpretation of events, but—if widely believed—a shaper of events. If that characterization is mistaken, it will misguide us. After a century vivid with the consequences of ethnic and racial hatred, and now deep into an era of weapons of mass destruction . . . the worst outcome would be for "clash" to become a self-fulfilling prophecy, intensifying conflicts or bringing about some that otherwise would not have occurred.

Stephen Walt feels much the same and gives his own warning:

If we treat all states that are part of some other "civilization" as intrinsically hostile, we are likely to create enemies that might otherwise by neutral or friendly. In fact, a civilizational approach to foreign policy is probably the surest way to get diverse foreign cultures to coordinate their actions and could even bring several civilizations together against us. . . . In this sense, The Clash of Civilizations *offers a dangerous, self-fulfilling prophecy: The more we believe it and make it the basis for action, the more likely it is to come true. Huntington would no doubt feel vindicated, but the rest of us would not be happy with the results.*

Richard Rubenstein, author of yet another critique, goes further by concluding:

Destructive conflict between identity groups, including pan-nationalist or civilizational groupings, can be averted and can be resolved if they do occur. But a violent clash of civilizations could well result from our continuing failure to transform the systems of inequality that make social life around the globe a struggle for individual and group survival—systems that feed the illusion that either one civilization or another must be dominant. . . . Huntington's call for the global defense of Western interests against competing civilizations therefore represents the worst sort of self-fulfilling prophecy.

Having reviewed scholarly critiques of "The Clash of Civilizations?" I turn to seeing it from the perspective of someone who has spent my life trying to bridge civilizations. First, the premise that civilizational interest can replace national self-interest is not supported by history. I find little evidence of a shift in direction away from nationalism to supranationalism and from state interest to "kinship" interest. When did states not act in their perceived self-interest? When will states no longer act in their perceived self-interest? The answer to both is "never." Self-interest guides the behavior of nation-states just as it guides the behavior of individuals. Artificially constructed civilizational groupings notwithstanding, the notion that Argentineans will somehow behave in tandem with Peruvians against the self-interest of Argentina is absurd. And more relevant to me, the notion that Pakistan will somehow behave against its self-interest because of some level of commonality with a monolithic and uniform interpretation of Islamic doctrine is implausible.

Contrary to the pontifications of many who are unashamedly contemptuous of Muslims around the world, democracy and Islam are congruent. The basic tenets of democratic governance are specifically and directly cited in Muslim teachings and are basic to the religion of Islam. As I have discussed, history shows that democracies do not make war against other democracies. And democracies are not state sponsors of terrorism. Therefore, I conclude (and challenge others to give any evidence to the contrary) that if democracies can be nurtured and sustained in the Islamic world, the possibility of conflict can be reduced. Indeed, the possibility of violent conflict between Islamic democratic states and Western democratic states, and the possibility of democratic state-sponsored terrorism by democratic Islamic states against Western targets, would be all but eliminated.

Carl Gershman, the president of the National Endowment for Democracy, has rightly pointed out that "deep-rooted historical and cultural resentment against the West" is being manipulated by what he calls a "pseudo-religious movement of ideological hatred, dictatorship and political dysfunction inevitably providing a breeding ground for terrorism." He adds that "the only long-term solution is to change the conditions that produce terrorism, which means encouraging the development . . . of successful democratic societies grounded in the rule of law and integrated into the global economy."

If the world community is to prevent a clash of civilizations, the way must be to promote the building blocks of democracy in the Islamic world. One prescription for peace is for democratic nations to promote democratic elections and democratic governance there. Islamic nations must be helped to create and sustain democratic infrastructure; to strengthen democratic elections and governance by supporting and training political parties; by assisting in the creation, operation, and funding of social welfare and human rights NGOs; by encouraging, supporting, and protecting a free press; by assisting in parliamentary and constitutional reforms; by helping to insulate an independent and autonomous judiciary from political manipulation; and by not rewarding military dictatorships but helping to equip militaries that accept the sovereignty of constitutional authority.

The building blocks of democracy, the infrastructure of democracy, are what, in the long run, will sustain democracy. Strengthening the institutions of democracy will give democracy a chance to succeed in all societies and especially in Islamic societies, where the stakes could not be higher. Promoting democracy promotes peace. Competitive political parties and functioning NGOs promote peace. A free press and an independent judiciary promote peace. A civilian-controlled military promotes peace. Even if the international community is reluctant to act out of moral impulses, it should act out of self-interest to build a more peaceful and stable world.

The entire concept of economic and cultural exchange, which is at the heart of public diplomacy throughout the West, is predicated on the belief—substantiated by evidence—that the more you know about others, the less likely you are to fear them. Indeed, one of the classic tools to promote understanding of democracy used by the West in the developing world has been programs to bring students to study in Western institutions. In the process they are not only educated in technology and the arts but are exposed to political norms and standards of individual empowerment that all human beings identify with and yearn for. It was in my years at Harvard University that I learned that political mobilization can not only impact public policy but also actually reverse bad policy. It was there that I learned that leaders can be held accountable to the people, and leadership that fails to address the popular sentiment can and will be replaced.

.

There is a group of nonacademics who approach the clash of civilizations thesis with positive prescriptions to reduce the violence they see in the world. They do not try to categorize this violence, necessarily, but seek to prevent it. Their only goal is to prevent violence between civilizations; to promote dialogue, cooperation, and concord. We shall call this group the reconciliationists. Most look to bring different groups around the world together for dialogue. These people believe that increased interaction and dialogue with and knowledge of the "other" will increase understanding and prevent conflict.

One proposal comes from an interesting source, the former (reformist) president of Iran, Mohammad Khatami, who proposed the theory of Dialogue among Civilizations in the forum of the U.N. General Assembly. Khatami believes that if civilizations get together and have dialogue, there will be more understanding in the world. Unlike the Huntington school, which sees only conflict in interaction, Khatami sees the sharing of ideas as a positive, binding force, a real step forward in intercivilizational interaction. He believes that nations have the responsibility to promote these dialogues. "Member states of the United Nations should endeavor to remove barriers from the way of dialogue among cultures and civilizations, and should abide by the basic precondition of dialogue. This fundamental principle rejects any imposition, and builds upon the premises that all parties to dialogue stand on essentially equal footing."

Khatami sees the spread of ideas across cultural barriers as inevitable, but without the spread of philosophy, religion, and the arts this spread of ideas is not complete and is dangerous, creating an identity crisis for people the world over. "Today, it is impossible to bar the transfer of cultural ideas among civilizations in various parts of the world."

However, the absence of dialogue among thinkers, scholars, intellectuals, and artists from various cultures and civilizations precipitates a dangerous "cultural homelessness": "Such a state of cultural homelessness deprives people of solace whether in their own culture or in the open horizon of World Culture."

As we have seen in many societies, such a void can be filled by emotional appeals, by irrationality, and by extremism. I am convinced that the authoritarian constraints on freedom of speech and of association placed by the Musharraf dictatorship in Pakistan are directly and causally related to the dramatic increase in extremism and violence in the North-West Frontier Province, in Baluchistan, and in the Federally Administered Tribal Areas. Indeed, if the intellectual choke collar of dictatorship is allowed to continue, there is a very real potential that this extremist effect will gather strength in the cities of Pakistan.

Khatami argued before the United Nations that a dialogue between cultures will not just help international relations but also help each culture better understand itself:

One goal of dialogue among cultures and civilizations is to recognize and to understand not only cultures and civilizations of others, but those of "one's own." We could know ourselves by taking a step away from ourselves and embarking on a journey away from self and homeland and eventually attaining a more profound appreciation of our true identity. It is only through immersion into another existential dimension that we could attain mediated and acquired knowledge of ourselves in addition to the immediate and direct knowledge of ourselves that we commonly possess. Through seeing others we attain a hitherto impossible knowledge of ourselves.

Finally, President Khatami warned that the dialogue among civilizations is difficult and requires political courage. But the possibilities are so great that it is definitely worth the effort. "Dialogue is not easy. Even more difficult is to prepare and open up vistas upon one's inner existence to others. Believing in dialogue paves the way for vivacious hope: the hope to live in a world permeated by virtue, humility and love, and not merely by the reign of economic indices and destructive weapons. Should the spirit of dialogue prevail, humanity, culture and civilization should prevail."

In response to these ideas by President Khatami, 2001 was declared by the United Nations to be the year of Dialogue among Civilizations. How ironic and horrible that the attacks of September 11 happened just days later. Nevertheless, a group of eminent scholars was chosen by the U.N. General Assembly to write a book on Dialogue among Civilizations that would explain the purpose and reasons for the dialogue:

> In the past, the perception of diversity as a threat was, and in some cases still is, at the very core of war. Ethnic cleansing, armed conflict or so-called religious clashes were all based on the perception that diversity is a threat. Recalling the atrocities of the previous decade, the answer to the question "Why do we need a dialogue?" seems simple and even obvious. The ancillary question then is: "Why now?" The process of globalization without dialogue may increase the probability and great danger of attempted hegemony. Diversity without dialogue may engender more exclusiveness. Therefore, a dialogue between those who perceive diversity as a threat and those who see it as a tool of betterment and growth is intrinsically necessary.

Tariq Ramadan is a Muslim living in Britain and the grandson of Hasan al-Banna, the founder of the Muslim Brotherhood. He would seem to be a man stuck between two worlds. He seeks reconciliation between Muslims and the West, yet some Muslims see him as selling out to the West, while the U.S. government won't give him a visa because it views him as supporting radical Islam. He believes that Islam needs to adapt to today's world: "The social and cultural norms of 8th-century Arabian Gulf tribes—some of which have been tightly woven into the practice of Islam over the centuries—are clearly irrelevant for European Muslims today."

Further, he forcefully argues for the flexibility of interpretation and the legitimacy of adaptability. He takes on the notion that Sharia cannot be changed. Sharia is not the rigid set of laws that some believe it is, he argues, but rather "Sharia is a set of values and principles, a path with objectives. Seeing it as a closed structure is a betrayal of its origins: When Muslims were confident, they took things from outside, but it is when you have no confidence that you define Islam in opposition to the West."

Ishtiaq Ahmed, a Swedish political scientist of Pakistani descent, adds a very strong economic point of view to the reconciliationist school. He understands that sustained economic inequality leads to the hopelessness, the feelings of inadequacy, and, ultimately, the hatred that feeds extremism and terrorism. "There is a clash between those forces which want to keep the material and intellectual wealth of the world confined only to the West and at most share it with corrupt elites in the Muslim world and those who struggle to universalize it for the benefit of all humankind."

When I entered Harvard University in the fall of 1969, Pakistan was under a military dictatorship. Although free elections were held in 1970 (the military believed a hung parliament would be elected), the mandate was not honored when two parties swept the two wings of East and West Pakistan.

In America I saw the power of the people to change and influence policies. While I was at university in America, my father was leading a movement in Pakistan for the political rights of the people of Pakistan. The struggle in Pakistan and the reality of the ability of people in America to assert themselves, to stand up without fear for what they believed in, were important influences in my life.

I was positioned between two worlds, the world of dictatorship and the world of democracy. I could see the power of the people in a democracy and contrast it to the lack of political power in my own country. I saw that people in America took their rights for granted: freedom of speech, freedom of association, freedom of movement. In my country people were killed or imprisoned fighting for these freedoms.

I was caught in the middle of two different worlds relating not only to dictatorship and democracy but also to gender. I was in America at an exciting time, when Kate Millett and Germaine Greer were writing books on women's rights, women's choices, and women's careers. My parents had taught me that men and women are equal in the eyes of God, that the first convert to Islam was a woman, that the Prophet of Islam married a career woman, that the line of the Prophet was carried through his beloved daughter Fatima, and that on the Day of Judgment all souls would be called in the name of the Mother.

But despite this emphasis on women's rights and the importance of women in Islam, all around me I could see that women were not treated with much importance in Pakistan, nor did they have many rights. Of course women had played an important part in Pakistan's independence movement. My mother had served as a captain in the National Guard. Fatima Jinnah, the sister of Pakistan's founder, and Mrs. Liaquat Ali Khan, the wife of Pakistan's first prime minister, had participated in the Freedom Movement. Yet women's rights were largely for the elite, with too few opportunities for women in general.

It was in America, over milk and cookies at night in my dormitory, that we would discuss how we wanted more out of life than the traditional roles of wife and mother. We believed that women should have the right to choose whether they want to live life as homemakers or seek careers.

In those formative years in the West I came to understand that the only limits on options for women in society are limits that we ourselves accept. I came to understand that by rejecting the constraints of the past on what women can do and become, we can build a newer world not only for ourselves and our daughters but for our husbands and sons. Just as democracy promotes moderation, gender equality promotes peace.

Our classes at Harvard were overflowing with undergraduates and graduate students from all over the world, especially from the developing world. Our exposure to life at Harvard and life in the United States empowered us and fundamentally changed our lives. When we returned to our homelands, whether to be educators, scientists, or prime ministers, our exposure to democratic values and institutions would ripple forth from us into our own societies. Just as democracy promotes peace, educational exchange promotes peace.

Technology and communication have changed our world and are influencing a global culture. The more one knows of people, the more comfortable one is with them. I believe that, even if Professor Huntington doesn't. Chatting on the Internet with strangers all over the world builds relationships and friendships and understanding. The ability to "Google" information from anywhere in the world puts technology into the hands of even the most isolated rural communities in the developing world. The more people learn, the more they want to learn. The more they interact, the less likely they will be to fear the unknown. Just as democracy and educational exchange promote peace, the free flow of modern technology and communication promote peace.

Everything we learn about sustained trade between nations tells us that it promotes understanding between cultures and civilizations. Globalization may be the most fundamental element of conflict resolution that has developed. The more nations trade with one another, the more they have to lose by engaging in conflict with one another. And we know that as individuals are exposed to more options in consumerism, in products they can purchase and use and share, the more they want options in other elements of their life.

It is one of the predicates of the free-trade movement that economic interaction bonds people. The European Community began as a common market that was designed to increase trade and break down barriers, to promote a sense of European commonality and decrease competition and tension between European states. In December 1988, during a summit retreat for the leaders of the South Asian Association of Regional Countries (SAARC), I proposed changing SAARC from a cultural association into an economic one. All the leaders enthusiastically endorsed the idea.

During my first term in office, the South Asian Preferential Tariff Agreement was born. In my second term it was ratified. It created the basis for economic transactions between the countries of SAARC.

In 1998, as Leader of the Opposition in Parliament, I proposed the creation of a free-trade zone in South Asia based on the European Common Market. We understand that the more interaction that takes place among people, states, companies, and universities in South Asia—and specifically Pakistan and India—the less likely it will be for these traditional enemies, now both nuclear-armed, to engage each other in potentially disastrous conflicts in the future. Just as democracy promotes peace, trade promotes peace.

Preventing a clash of civilizations, at least in terms of the Islamic world, requires that we put our trust in the power of trade, exchange, technology, education, and democratic values to accelerate the process by which Islamic societies can build bonds and trust with Western societies. It is by externalizing exposure that we learn to tolerate and respect the values of other civilizations. It is not by insulation but by interaction that the world will become safer. By increasing economic, social, and political development with the Islamic world, the West and East can protect common and universal values true to the Abrahamaic heritage.

Finally, to understand different civilizations, we need to understand the fault lines that could lead to conflict. We are dealing not with a clash between civilizations but rather with a clash within a civilization. The most critical battle for the hearts and souls of the successor generation of Muslim leaders, and for the passion of the Muslims around the world, is not a battle with the West. The debate is between different interpretations of Islam, different visions for the

Muslim Ummah. It is about the lack of tolerance that some interpretations show for other interpretations within Islam or other religions outside Islam.

It is the value of tolerance that will be the deciding factor between the forces of extremism and the forces of moderation, between the forces of dictatorship and the forces of democracy, between fanaticism and education, between pluralism and bigotry, between gender equality and gender subservience, between inflexible traditionalism and adaptable modernity.

In other words, the real clash within and outside Islam is a battle between the past and the future. It is the resolution of this battle that will determine the direction not only of the relationship between Islam and the West but of international relations in this century. Without further delay, to break the chains and cycle of poverty, extremism, dictatorship, and terrorism, we need to move on the path toward true reconciliation.

6

Reconciliation

There are two historic clashes unfolding in the world today that appear inexorably intertwined. The resolution of one could determine the immediacy of the other. The internal clash within the Muslim world is not merely over theology. The real fight is not over the succession to the Holy Prophet that divides the Shiite and Sunni communities. It is certainly not about the language of the Holy Quran. It is not really about the interpretations of Sharia. The extremism and militancy of Muslim-on-Muslim violence is a long battle for the heart and soul of the future not only of a religion but also of the one billion people who practice it. Fundamentally, it is also about whether the Muslim people can survive and prosper in the modern era or whether linkages with traditional interpretations of the sixteenth century will freeze them in the past. If Muslims can adjust to changes in the political, social, and economic environment we will not only survive but flourish. If modernity is dogmatically resisted, the existence of Muslims as a viable community will become vulnerable. In the extreme, Muslims will attempt to impose themselves in a messianic union of Muslim states that could provoke the external clash between Islam and the West that the world is focusing on today.

There is much that Muslims can do to reconcile the internal contradictions that badly divide their communities in the twenty-first century. By charting such a course, Muslims can again become one of the central forces shaping the future of humanity. The West too can bridge the gap between itself and the Muslim community by taking specific and concrete steps. This widening gap of perception, values, and sense of compatibility is what threatens to explode into the epic battle of the twenty-first century. The Islamic states, in my view, can both accommodate and reconcile with one another and with the West. It is an ambitious undertaking, but it can be done.

.

Muslim scholars and leaders have bemoaned the community's loss of power—political, intellectual, scientific, and economic—since the colonial era. Although Westerners are not fully aware of its dimensions, an important debate has raged among Muslims over how to deal with modernity. Just as some have called for rejecting modern ideas, and in the most extreme cases have advocated an endless war with the West as the source of modernity, others have proposed strategies for reconciling the Islamic world with modern scientific ideas and with the modern political, economic, and social environment.

At the beginning of the twentieth century, reformist ideas appealed to the Muslim intelligentsia. Even unlettered farmers paid attention to speeches and poems by Islamic reformers stressing the need for improving the fortunes and influence of Muslims through mass education, democracy, and economic progress. But then, tragically, most of the Muslim world fell under the sway of dictatorial regimes. Irrespective of whether the dictators espoused secular or religious ideas, the stifling of debate undermined the pluralist environment necessary for an Islamic reformation. Dictatorship choked the oxygen of innovation.

Among the earliest voices for modernist reform in the Muslim world was the poet-philosopher Mohammad Iqbal (1877–1938). He is also considered the spiritual father of Pakistan, even though he died almost a decade before Pakistan's creation. In his book *Reconstruction of Religious Thought in Islam,* Iqbal questioned the rationale for the failure to reinterpret the principles of faith, unmodified since the twelfth century. He called for looking beyond the traditional Islamic schools of jurisprudence and reviving *ijtihad*. *Ijtihad* means "reason" and Islam calls upon Muslims to apply reason in seeking a solution to their problems. According to Iqbal:

> *The law revealed by the Prophet takes special notice of the habits, ways and peculiarities of the people to whom he is specifically sent. The sharia values {ahkam} are in a sense specific to that people; and since their observance is not an end in itself they cannot be strictly enforced in the case of future generations. . . . Let the Muslim of today appreciate his position, reconstruct his social life in the light of ultimate principles, and evolve, out of the hitherto partially revealed purpose of Islam, that spiritual democracy which is the ultimate aim of Islam.*

The contemporary Pakistani academic and Professor Fazlur Rehman echoed Iqbal's philosophy. He wrote:

> *The Muslims' aim of Islamizing the several fields of learning cannot be really fulfilled unless Muslims effectively perform the intellectual task of elaborating an Islamic metaphysics on the basis of the Quran . . . the effort to inculcate an Islamic character in young students is not likely to succeed if the higher fields of learning remain completely secular, that is unpurposeful with regard to their effect on the future of mankind. . . .*
>
> *And here appears the stark contrast between the actual Muslim attitudes and the demands of the Quran. The Quran sets a very high value on knowledge. . . . By contrast the Muslim attitude to knowledge in the later medieval centuries is so negative that if one puts it beside the Quran one cannot help being appalled. . . .*
>
> *It is the biographers of the Prophet, the Hadith collectors, the historians, and the Quran commentators who have preserved for us the general social historical background of the Quran and the Prophet's activity and in particular the background of the particular passages of the Quran. This would surely not have been done but for their strong belief*

*that this background is necessary for our understanding of the Quran.
It is strange, however, that no systematic attempt has ever been made to
understand the Quran in the order in which it was revealed, that is by
setting the specific cases of the* shu'un al-nuzul *or "occasions of revelation" in some order in the general background that is other than the
activity of the Prophet (the sunna in the proper sense) and its social environment. . . .*

*The new step consists in studying the Quran in its total and specific
background (and doing this study systematically in a historical order),
not just studying it verse by verse or passage by passage with an isolated
"occasion of revelation."*

Nurcholish Majdid, a prominent Indonesian academic and writer
who died in 2005, pressed for review, reinvention, and reinvigoration
of Muslim scholarship and theology. Majdid called for the need for
deep-seated changes in Islam to keep up with changes in the world:

*. . . renewal has to start with two closely related actions, that is, freeing
oneself from traditional values and seeking values which are oriented
towards the future. The orientation to the past and excessive nostalgia
have to be replaced by a forward-looking attitude. . . . What has happened is that the Ummah has lost its creativity in this temporal life to
the extent that it leaves the impression that it has decided not to act
rather than risk making a mistake. In other words it has lost the spirit
of Ijtihad. Therefore Ijtihad or renewal ought to be a continuous process
of original thinking based on the evaluation of social and historical
phenomena which, from time to time, need to be reviewed in order to
determine whether they are really erroneous. . . .*

*Iman, Taqwa and the religious experience of the appreciation of di-
vinity, as explained above, are spiritual in nature. This means that
these values are connected to something that is entirely intrinsic in a
person, namely sincerity. Hence they have to grow out of free and independent choice. For this reason, religious values are highly individual
in nature.*

Abdul Karim Soroush, an Iranian pharmacologist and philoso-pher, argued for a rethinking of Islam's relationship with the non-Muslim West. He believed that Islamic nations must allow and encourage advances in science. He believed that "revealed religion is divine but the science of religion is a thoroughly human production and construction."

Text does not stand alone, it does not carry its own meaning on its shoulders, it needs to be situated in a context, it is theory-laden, its interpretation is in flux, and presuppositions are as actively at work here as elsewhere in the field of understanding. Religious texts are no exception. . . .

Revealed religion itself may be true and free from contradictions, but the science of religion is not necessarily so. Religion may be perfect or comprehensive but not so for the science of religion. Religion is divine, but its interpretation is thoroughly human and this-worldly. . . .

The whole history of the sincere efforts of commentators to liberate Quranic commentary from the infiltration of external ideas has ended in one sharp and important result, namely, the practical unavoidabil-ity of such infiltration, together with its epistemological inevitability.

Dr. Mohammed Arkkoun, an Algerian who received his doctorate at the Sorbonne, attempted to reshape Islamic interpretation by using contemporary social-scientific methods: "There is a need to encourage and initiate audacious, free, productive thinking on Islam today."

Religious tradition is one of the major problems we should rethink to-day. First, religions are mythical, symbolic, ritualistic ways of being, thinking, and knowing. They were conceived in and addressed to societ-ies still dominated by oral and not written cultures. Scriptural religions based on a revealed Book contributed to a decisive change with far-reaching effects on the nature and functions of religion itself. Christian-ity and Islam (more than Judaism, until the creation of the Israeli state) became official ideologies used by a centralizing state which cre-

ated written historiography and archives. There is no possibility today of rethinking any religious tradition without making a careful distinction between the mythical dimension linked to oral cultures and the official ideological functions of the religion. . . .

We must rethink the whole question of the nature and the functions of religion through the traditional theory of divine origin and the modern secular explanation of religion as a social historical production. . . . We need to create an intellectual and cultural framework in which all historical, sociological, anthropological, and psychological presentations of revealed religions could be integrated into a system of thought and evolving knowledge. We cannot abandon the problem of revelation as irrelevant to human and social studies and let it be monopolized by theological speculation.

K. H. Abdurrahman Wahid, the former president of Indonesia, questioned the inflexibility of rigid Islamic interpretations applied to the modern political environment, stressing Islamic tolerance:

All too many Muslims fail to grasp Islam, which teaches one to be lenient towards others and to understand their value systems, knowing that these are tolerated by Islam as a religion. The essence of Islam is encapsulated in the words of the Quran, "For you, your religion; for me, my religion." That is the essence of tolerance.

Religious fanatics—either purposely or out of ignorance—pervert Islam into a dogma of intolerance, hatred and bloodshed. They justify their brutality with slogans such as "Islam is above everything else." They seek to intimidate and subdue anyone who does not share their extremist views, regardless of nationality or religion. While a few are quick to shed blood themselves, countless millions of others sympathize with their violent actions, or join in the complicity of silence. This crisis of misunderstanding—of Islam by Muslims themselves—is compounded by the failure of governments, people of other faiths, and the majority of well-intentioned Muslims to resist, isolate and discredit this dangerous ideology. The crisis thus afflicts Muslims and non-

Muslims alike, with tragic consequences. Failure to understand the true nature of Islam permits the continued radicalization of Muslims world-wide, while blinding the rest of humanity to a solution which hides in plain sight. The most effective way to overcome Islamist extremism is to explain what Islam truly is to Muslims and non-Muslims alike. Without that explanation, people will tend to accept the unrefuted extremist view—further radicalizing Muslims, and turning the rest of the world against Islam itself. Muslims themselves can and must propagate an understanding of the "right" Islam, and thereby discredit extremist ideology. Yet to accomplish this task requires the understanding and support of like-minded individuals, organizations and governments throughout the world. Our goal must be to illuminate the hearts and minds of humanity, and offer a compelling alternate vision of Islam, one that banishes the fanatical ideology of hatred to the darkness from which it emerged.

Rather than view Islam and its teachings as a single monolithic entity, it is more accurate to recognize and acknowledge the de facto plurality of opinions that have always existed as to what Islam is, and what it "compels" you, me or us to do. A vast diversity of opinions about Islam are held by my fellow Muslims, which they are free, in turn, to share with me. . . . For example, those wishing to "purify" Islam from so-called bid'a, *or innovation, may reject the use of a drum to issue the call to prayer, reverence of saints, or even the use of a rosary while reciting the names of God. Thus we may refer to others' personal experience and understanding of Islam as "Your Islam," and go through life adopting or politely refusing to adopt any given element thereof. . . . Such sharing of views may or may not produce what we might call "Our Islam," dependent on the respective understanding and experiences of those involved, but at least it fosters mutual respect and tolerance of differences. For me personally, "Their Islam" is a fair term to describe the views of those who would annihilate the great beauty and diversity of traditional Islam in the name of an artificial and enforced conformity to their own rigid opinions. For such people, Our Islam is a misnomer, for in fact they seek to enforce—through intimidation and violence—a colorless, monolithic uniformity that does not and has never existed in the long history of Islam.*

The Indian Muslim scholar and writer Maulana Wahiduddin Khan endorses this view of Islamic pluralism:

Once, when Prophet Mohammad was sitting with his companions at Kabbah, a poetess called Hind approached him and recited a poem, which said: "Mohammad is a condemned person and we deny him." The meaning of Mohammad is "one who is praiseworthy"; but she called him "Muzammam," which means "the condemned one." The Prophet just smiled. His companion, Abu Bakr, asked how he could smile at such derogatory language. The Prophet replied,

"My name, given to me by my family, is Mohammad, while she talks of Muzammam; so all the curses apply to him and not me. Why should I bother then?" For him, the slight was too trivial to merit any reaction.

Muslims mistakenly regard it as their duty to stop any visual depiction of Prophet Mohammad. This is untrue. It is the followers of Islam who are forbidden to do so in order to discourage idolatry. Moreover, Islam forbids imposing its beliefs on people of other faiths. Even in Muslim countries, Muslims cannot impose their laws or culture on others.

Islam's mission is to spread the message of peace, love, tolerance and character-building, so that people may live in accordance with the will of God. To do so effectively, the relationship between Muslims and non-Muslims must be smooth. Muslims are only defeating their own purpose by creating an environment of hate and mistrust. The Muslims are playing into the hands of the divisive forces with their violent protests. By increasing anti-Muslim feelings, they are uniting the world against themselves. Instead, they should aim towards normalcy.

The Pakistani scholar Muhammad Khalid Masud emphasizes Islamic pluralism and compatibility with other religions:

We may say that Islam's encounter with the West began quite early when Greek intellectual tradition came to the Muslim society in the ninth century under the Abbasid caliph Mamun. It generated a debate about the legitimacy of studying Greek sciences but the fact that a movement in favor of Greek sciences continued to flourish throughout the Islamic history is by itself an evidence of Islam's cultural compatibility. This movement generated a controversy deep into the religious issues about the nature of the Quranic revelation and the role of reason. Opposition to this movement finally triumphed as "orthodoxy," but not without recognizing the necessity of studying Greek secular sciences. The science of Kalam *(roughly, Islamic theology) born out of the need to refute foreign ideas was regarded as the science of* Da'wa *and Jihad. This science relied heavily on Greek logic, rhetoric and metaphysics. . . .*

No doubt, opposition in the Muslim community to rational sciences continued, but at the same time a considerable number of Greek texts or their Arabic adaptations on Astronomy, Geography, Calculus, Mechanics etc., became part of the Muslim religious curriculum. They were called "Ma'qulat" (rational sciences) in contrast to pure religious text studies, which were called "Manqulat."

Islam as a moral tradition favors pluralism on two bases. Firstly because it appeals to human reason. . . . The second basis of pluralism is social acceptance of these values. This basis also regulates the dissent. The Qur'an calls good Ma'ruf (well known) and evil Munkar (rejected), which points to the fact that normativity is based on social acceptance or rejection. The social dialectics develop the acceptable definition of ethical values.

The reformist thinkers cited include traditionally educated theologians. Their views cannot be rejected out of hand by other ulema as a deviation from the Islamic path. The existence of a vocal group of reform-minded Islamic scholars could have prevented the drift into extremism by significant segments of the Muslim world. But the absence of democracy helped extremists against the Islamic reformists.

In several Muslim states, dictatorships tended to favor hard-line traditionalist interpretations of religion in return for the theocrats' providing a fig leaf of religious legitimacy to autocracy. Secular dictators repressed religious debate altogether. This benefited the extremists because they organized underground, whereas the moderate reformers were unable to do so. In recent years, the rise of Islamist extremism and the threat of terrorism have resulted in intimidation or silencing of reformist voices or caused them to be lost in the din of violent messages. Democratic political systems would provide protection to reformist and modernizing theologians and philosophers. The ideas of Mohammad Iqbal, Fazlur Rehman, and Nurcholish Madjid could then be taught in schools to a younger generation growing up in a society where free exchange of ideas trumped any dictator's demand for conformity. And the living reformers like Muhammad Arkoun, Abdur Rehman Wahid, Wahiduddin Khan, and Khalid Masud would be able to preach and teach their modernizing theology without facing repression or marginalization by the state.

·

In making the case that much of the Muslim world's future depends on whether democracy can replace authoritarianism and dictatorship, my premise is that democracy weakens the forces of extremism and militancy. And if extremism and militancy are defeated, our planet can avoid the cataclysmic battle that pessimists predict is inevitable. Thus much of what I think needs doing to defeat Islamic extremism centers around what I think must be done to strengthen democracy among Islamic states. I believe we must outline what all societies— but specifically Islamic societies—can do to give democratic governance a chance to succeed.

·

Democracy cannot be sustained around the world in the absence of a stable and growing middle class. Huge economic disparities between social classes in a society strain national unity, creating a gap between the rich and the poor. Educated and rich "elites" dominating illiterate masses are not a successful prescription for building a democratic society. Take the case of Costa Rica in the context of the rest of Central America and South America. Costa Rica is distinguished from its neighbors by the size, stability, and influence of its middle class. Nicaragua, Panama, and El Salvador witnessed a progression of autocrats and dictators for decades, while Costa Rica remained democratic and politically stable. I believe that the existence of a middle class is the key element to democratic infrastructure.

Some may point to India as an exception to the generalization I am making, but is it? Indian democracy stabilized and its caste system contracted as its middle class expanded. The growth of India into a regional and international economic power occurred—not coincidentally—as its middle class exploded into a huge economic and political force. The entrepreneurial mind-set of the Indian middle class has been at the vanguard of India's technological leap forward. As the middle class grew and expanded, the social limitations constraining India in the past, especially its inflexible social class system, softened, and they will eventually disappear. The end of the caste system will herald a true democratic era for the Indian nation, demonstrating that democracy and development go together.

·

But how can a nation build a middle class? The first key is to build an educational system that allows children to rise to a higher social and economic status than their parents—in other words, an educational system that delivers hope and real opportunity is a prerequisite for democracy. Good public educational opportunity is the key to the economic and political progress of nations, and it can be so in the Islamic world as well. Building a strong, compulsory educational system requires two key elements. First, compulsory public education—for all citizens, all classes, and both sexes—must be a priority. But one needs more than the will to make it a priority. One also must have the means. It is essential that budgets for Muslim countries be prioritized by social need, not outdated political or military history. In Pakistan, for example, $4.5 billion is spent on the military each year. This is an astounding 1,400 percent more than is spent on education! Military versus social sector foreign assistance is even more disproportionate. Pakistan has a strong military with plenty of tanks and missiles, but it lacks a dynamic and technologically educated workforce. The key to investing in the future is to invest in people's educational opportunities.

As prime minister, I attempted to put as much funding into the social sector and education as I could. Overburdened with the debts run up by dictatorship, my government still built almost fifty thousand elementary and secondary schools around the country, and especially in the rural areas. I wish our debts had been rescheduled so we could have done more. The fundamental constraint upon my governments in prioritizing our budget was the enormous percentage of our GNP that was diverted to debt repayment and defense.

To complicate matters, the military came under the president and not the prime minister. It was difficult to ask for more transparent accounting of the huge funds made available to the military without constitutional authority. Moreover, as in many developing countries, the military was an institution that had been insulated from civilian control and direction for decades under one military dictatorship or another.

From the tenuous fortnight between my party's victory in November 1988 and the time I formed a government, I was under enormous pressure—from the public, the military, and key international players—all of whom expected a chunk of the federal budget, which was already burdened with debt. All this occurred while international financial institutions, including the International Monetary Fund, were pressing me to cut national expenditure to reduce the budget deficit. This undermined my ability to govern effectively. Such pressures constrained me, as they have other leaders in new democracies. If education is to succeed in a nation like Pakistan, or the Islamic world and developing world, new democratic leaders need the international and political support to withstand militaries' destabilizing them with ambitious generals keen to rule once again. Armies should protect borders, not rig elections or blackmail elected leaders. And a military that is subservient to civilian rule would strengthen the ability of democratic institutions to take hold. Democratic governance can take place when governments are safe from the sword of Damocles of military takeover constantly swinging over Parliament's head. Often, when the military does leave government, it leaves behind a constitution in which power is divided between the president and the Parliament. The Parliament is the voice of the people. The president becomes the voice of the military. In the clash the people are the casualties.

Another important way in which education can build democratic infrastructure in the Islamic world concerns the real threat from militant madrassas. Many of the madrassas across Pakistan and other parts of the Islamic world make a significant contribution to education for the nation that is not dissimilar to that of parochial educational institutions in the West. The great threat to democratic governance is not the genuine educational madrassas but rather the militant madrassas. These political and military training camps invest little time and resources in primary education. Rather, they manipulate religion to brainwash children into becoming soldiers of an irregular army. They conduct hours upon hours of paramilitary training. They teach hatred and violence. They breed terrorists, not scientists. Militant madrassas undermine the very concept of national identity and rule of law. When I was prime minister, I invested enormous political resources in stopping these paramilitary political madrassas, but unfortunately, in the twelve years since I left office, these militant madrassas have spread like a forest fire. We believe that as many as twenty thousand new ones have been built in Pakistan alone during this time.

These militant madrassas did not flourish because Pakistani citizens suddenly became more religiously orthodox than ever before in our history. They took advantage of parents from low-income social classes who wanted a better life for their children. If parents are so poor that they cannot house, clothe, feed, and provide health care for their children, and the state fails to provide such basic human needs through public services, they will seek an alternative. The militant madrassas have become, over time, an alternative government for millions of Pakistanis. The destabilization of democracy has led to the success of the militant madrassas.

The militant madrassas provide a social service delivery system that appeals to the poor. That is what empowered the great political machines of America's cities in the early part of the twentieth century. Not only do the children in the militant madrassas become recruits for a successor generation of militants and extremists, but their parents become dependent on the madrassas for responding to their social needs. This creates a symbiotic relationship that nourishes extremism and breaks down the writ of the government. Militant madrassas are dangerous to all societies. They should be stopped, not just in Pakistan but all over the world where they produce the child soldier. If a viable state educational alternative existed that would provide both education and social services to the children of the poor, the militant madrassas, breeding grounds of violence, would shrivel and dry up.

·

The next fundamental change needed within Islamic states to equalize society and opportunity deals with women's rights. In any society, gender equality is a prerequisite for democracy to thrive. This is especially true in Islamic societies, where gender inequality has been used to promote political subordination and domination for centuries. It stifles social growth and opportunity. Societies with gender equality have without exception been pluralistic, tolerant, economically viable, and democratically stable.

As a person growing up in an environment of gender equality—an environment in which daughters and sons were treated equally—I have found it difficult to tolerate gender inequality in any form. I find it offensive both as a woman and as a Muslim. In 1997 the Taliban shut down girls' schools in Afghanistan and kept women off the streets. In Pakistani territory now ceded by the Musharraf administration to the Taliban and Al Qaeda, girls' schools are being shut down—sometimes even burned—and women stripped of their constitutional rights. The law must be gender-blind. Democracy cannot work if women are subjugated, uneducated, and unable to be independent. I worked hard as prime minister to eradicate illiteracy among grown women in Pakistani society for reasons transcending morality or justice. It is known that literate mothers raise literate children. One of the most efficient ways to dent illiteracy in society is to educate mothers. Islamic societies that fail to educate women condemn their children to a vicious cycle of ignorance and poverty. From illiteracy and poverty stem hopelessness. And from hopelessness come desperation and extremism.

.

An important way in which women's rights, economic development, and the building of a middle class come together is the economic empowerment of women. My father encouraged his daughters to be as well educated as our brothers and also to be economically independent. A true measure of liberation from traditional roles and traditional subordination by men is the extent to which women are economically self-sufficient. If the Prophet's wife could work outside the home, all Muslim women should be free to work. Economic independence brings political independence, and political independence within the family encourages pluralism and democratic expression and organization outside the family.

This phenomenon emerged more than two decades ago in the brilliant innovation of the Grameen Bank in Bangladesh, which won the Nobel Peace Prize in 2006. The concept, born in the village of Jobra, was to encourage and facilitate women to start small businesses through "microfinance." A one-village experiment transformed into a full-fledged banking operation providing small loans to those who had never before been able to receive financial credit. These micro-loans did not require legal instruments as collaterals. The goal was poverty reduction. The founders recognized poverty as a threat to democracy and to international peace. Ninety-seven percent of the recipients were women, and the recovery rate was above 98 percent. The bank, now with approximately 2,500 branches in 80,000 villages, has distributed almost $7 billion U.S. in loans, trying to jump-start an entrepreneurial class among the women of Bangladesh. My government introduced the concept of microcredit in 1989 in Pakistan. But the repeated dismissals of our government did not allow this concept to flourish as it should have.

The concept of development banks to create an entrepreneurial class is widespread in the developing world, although microcredit is a relatively new phenomenon. My father established development banks in Pakistan in the 1970s. There is an Asian Development Bank. But microcredit banks are something new. China has turned to it, too, opening a Chinese Development Bank that provides microcredit and small lending for business development. Working off the theme of microcredit to eradicate poverty in Pakistan, my government came up with the idea of a women's bank to give credit to women, thereby opening opportunities for them. As prime minister I started a Women's Development Bank, which provided start-up loans to women in Pakistan (we allowed men to deposit money in this bank, but loans were targeted to women only!). The expansion of micro-finance within the Islamic world can encourage the creation of an economically and politically stable middle class. Special funds reserved for microloans and small loans to women will encourage entrepreneurship as well as economic and political independence.

·

Political and social reforms are often interrelated. In the fall of 2007, a woman victim of gang rape in Saudi Arabia was sentenced to sixty lashes and six months in jail. This shocking incident drew dismay and disbelief around the world and triggered shock waves within

Saudi society. It acted as a catalyst for Saudis to reexamine basic doctrine perpetuated in society in the name of religion without theological foundation. It caused the king to pardon the victim. It caused Muslims around the world to reexamine traditional interpretations of religion that inflict injustice and freeze social development. The Saudi incident served to reinforce yet another basic recommendation necessary to reconcile the clash within Islamic societies. Women's rights groups have been at the vanguard of the fight for human rights and building a viable civil society. The development of a strong civil society is a basic building block of democracy. Nongovernmental organizations that deal with women's rights, human rights, and the rule of law are key to democracy. There can be no democracy without a stable and protected civil society. The fact that General Musharraf, in the first moments of his second declaration of martial law in November 2007, arrested thousands of activists, lawyers, and judges, demonstrated that he knew full well that a thriving civil society is incompatible with dictatorship.

Civil society is a concept intrinsically linked to strong democratic traditions, giving real meaning to the concept of pluralism in society. Nongovernmental groups, community organizations, women's organizations, student unions, trade unions, environmental organizations, professional associations, and religious groups each represent the interests of particular constituents. Collectively, they form the foundation of democracy in theory and practice. The groups making up civil society are often at the vanguard of political reform and demands for governmental transparency. They are the internal election monitors. They stand up against violations of human rights. They work with international groups that promote democracy to guarantee a fair political process but not a guaranteed political outcome.

Such civil society groups can be both powerful and credible. Although civil society cannot replace political parties in the democratic process, it complements political parties by ensuring a level playing field in politics. By working with their counterparts around the world, NGOs in the Muslim world can integrate societies and break down walls of ignorance. Civil society is therefore invaluable to building democratic systems that isolate extremists. There can be no democracy without a civil society, just as there can be no democracy without the rule of law. Laws should be enacted to protect civil society from political attack. Financial and tax incentives can further strengthen civil society institutions. Here again, women's organizations have played the most crucial role in civil society in promoting political reform in the democratically developing world.

·

We need a powerful, heavily networked international group aggregating activist women's groups throughout the Muslim world to create, in the title of Professor Wadud Amina's book and conference, an "international gender Jihad for women's rights." The creation of Women of the Ummah (WOTU), an organization aggregating women's rights groups throughout the Muslim world, can help Muslim women to act as catalysts for a democratic society that challenges the very dictatorship that breeds extremism.

Women's groups can serve as the linchpin of civil societies around the world. Special attention should be paid to organizing women as political, social, and economic players in each respective society. This is especially true for the Islamic world, in which women often face subjugation. This subjugation has come not from the message of Islam, which proclaims the equality of men and women, but from narrow interpretations of Sharia that deliberately promote subjugation and from political exploitation by ideological clerics. The prohibition on women driving cars is one such example. Automobiles did not exist when Islam dawned in A.D. 612. Therefore the prohibition on women driving cars is all the stranger. Such narrow interpretations of Islamic scripture—usually written by traditional men from traditional societies to underpin traditional authority—need to be revisited in the light of Islamic principles.

Again, the denial of education that the Taliban practiced has no foundation in Islamic law. The exploitation of women has had a dev-

astating ripple effect across Islamic society and across Islamic generations. Political reform, social reform, and economic reform for women
are tied together; one cannot proceed without the other. Women in
Islamic society can serve as a catalyst for reform across the political,
electoral, and government sectors. To effect this change, women must
be organized. Thus it is critical that women's groups expand and that
women's groups throughout Muslim societies join together to seek
common changes in their societies. Some international Islamic wom-
en's groups do already exist, but they are relatively weak and un-
funded and lack political clout. Among them are the International
Federation of Women Against Fundamentalism and for Equality, the
American Muslim Women's Association, the United Muslim Women
Association, the South Asian Women's NETwork, the Sufi Women
Organization, the Federation of Muslim Women, and BAOBAB for
Women's Human Rights.

Women, actively organizing over the last several years, are becoming potent political actors within their own countries in the Muslim
world. I have selected representative Muslim countries—Pakistan,
Jordan, and Turkey—and have been impressed by the sophistication
of the women's rights networks that exist in those countries.

·

Here is a sampling of women's rights groups working in Pakistan:
Learning & Caring Society is an organization dedicated to training
and teaching women, and the mentally disabled, skills aimed at improving their socioeconomic status. Pakistan Association for Women's
Studies is a social welfare organization that works to provide a forum
for interaction and coordination for those engaged in teaching, research, or action for women's development. ROZAN is an Islamabad-
based nongovernment organization (NGO) working in Pakistan on
issues related to emotional and psychological health, gender, and
violence against women and children. The Pakistan Women Lawyers
Association is an NGO delivering professional legal services and economic counseling to women. Women Citizen Community Board tar-

gets the rural women of Kasur, holding seminars on women's rights and education. All Pakistan Women's Association for Health and Economic Welfare is a nonpolitical organization whose fundamental aim is to safeguard the moral, social, and economic welfare of women and children in Pakistan. The Aurat Foundation is a civil society organization committed to women's empowerment in society. War Against Rape is a group of committed women and men dedicated to building a sensitized society free from gender-based oppression, discrimination, exploitation, and violence. Women-alert is an online pressure group of women activists dealing with issues of sexual harassment and gender discrimination of women in public and in the workplace.

.

Jordan Forum for Business and Professional Women provides a platform that develops, empowers, and advocates for business and professional women. Jordanian Women's Union undertakes programs on income generation and skills training courses for women. Young Muslim Women's Association for Special Education offers educational opportunities to mentally challenged children, children with learning difficulties. Jordanian National Committee for Women is the authority on women's issues and activities in Jordan's public sector.

.

Istanbul Bar Association Women's Rights Enforcement Center is an organization founded to enforce women's rights. Willows Foundation is an NGO begun in 1998 to make Turkish women aware of their reproductive rights and have access to accurate information about their rights. National Council of Turkish Women combines women's groups in Turkey. Purple Roof Women's Shelter and Foundation is a group that runs women's shelters and counseling centers. Women's Library and Information Centre Foundation is the first and only women's library and information center in Turkey.

.

Extremism, militancy, terrorism, and dictatorship feed off one another, thriving in an environment of poverty, hopelessness, and economic disparity among social classes. This symbiotic relationship of extremism, militancy, terrorism, dictatorship, and poverty is a direct threat to international and national stability and a clear danger to world peace.

Targeted economic development can help reduce poverty and violence in Muslim-majority states. Alleviating poverty is a fundamental responsibility of all Muslims, wherever they live, as part of the basic principles of Islam. It would be far more Islamic in its true sense to declare a jihad on poverty, illiteracy, hunger, and poor governance. That is exactly what I am proposing.

Islam's first generations produced knowledge and wealth that empowered Muslim empires to rule much of the world. But now almost half the world's Muslims are illiterate. The combined GDP of the member states of the Organization of the Islamic Conference (OIC) is about the same as that of France, a single European country. More books are translated annually from other languages into Spanish than have been translated into Arabic over the past one hundred years. The 15 million citizens of tiny Greece buy more books annually than do all Arabs put together.

The World Bank comparison of average incomes demonstrates a disquieting pattern. In the United States, the average per capita income is almost $38,000; in Israel it is almost $20,000. Pakistan, on the other hand, has an annual per capita income that barely crosses the $2,000 mark. No Muslim nation that is a non–oil producer has an annual per capita income near or above the world average. I find this pattern, these statistics, unacceptable.

·

The chain must be broken. One direct way to do that would be for the Gulf states to jump-start economic and intellectual development in the rest of the Islamic world. This is what my father tried to do for Pakistan in the 1970s, and this is what I tried to do as prime minister in my two terms in office. Norway and Kazakhstan, as examples, provide models of committing oil revenues to internal economic development and foreign investment that can be refined to address the economic, social, and political realities of the non-Gulf Islamic world. In other words, oil can break the chain of poverty, hopelessness, dictatorship, and extremism that often ruptures into international terrorism.

The Alaska Permanent Fund is Alaska's way of equitably distributing income the state acquires from its mineral wealth. In 1969 the Prudhoe Bay oil lease sale yielded $900 million to the state of Alaska. This income was about eight times as much as the entire state budget of Alaska in 1968, $112 million. The money was earmarked for important projects, but money can run out. Thus in 1976 an Alaskan constitutional amendment was passed establishing the Alaska Permanent Fund. Under the amendment, "at least 25 percent of all mineral lease rentals, royalties, royalty sales proceeds, federal mineral revenue-sharing payments and bonuses received by the state are placed in a permanent fund, the principal of which may only be used for income-producing investments." The Alaskan fund now earns money for the Alaskan state by investing 25 percent of its income in economic projects.

This Alaskan fund invests in both high- and low-income investments. The principal cannot be spent but only reinvested. The fund earnings are spent on various state projects, but the majority of the earnings are given to the Alaskan people in the form of an annual dividend. The dividend varies from year to year, but it averages between $600 and $1,500 per person. Every man, woman, and child who is an Alaskan resident for a minimum of one year is eligible to receive the yearly dividend.

The Government Pension Fund of Norway, established in 1990, is somewhat different from the Alaskan model. The Norwegian government places taxes from oil and gas companies, as well as licenses granted to these companies, in a separate fund. This account is the largest pension fund in all of Europe. The fund invests globally as well

as domestically. Given Norway's small population, the country determined from the outset that the fund could not be invested solely in
the country's own economy, as the economy could not absorb that level
of stimulation. The fund is used to cover years when the Norwegian
government budget is in a deficit. Its basic purpose is to save for future
generations, to invest in the economy, and to pay public pensions.

Kazakhstan gets a huge part of its annual revenues from the development of Caspian Sea oil deposits. In August 2000 President Nazarbayev established a National Fund. The concept was inspired by the
Alaska Permanent Fund and partially modeled on the Norwegian
fund. The Kazakhstan National Fund obtains income from the taxes
and payments made by raw-mineral companies to the government.
The fund invests heavily in foreign markets but is also used to stimulate industry within Kazakhstan. Unlike the Alaskan and Norwegian
funds, the goals of the fund and the distribution of proceeds are not
clearly specified.

Community responsibility is an inherent part of Islam. The Islamic world is fortunate in that part of it is floating in oil revenues.
However, countries such as Pakistan, Bangladesh, and Indonesia
struggle to provide basic human services to their populations. In verse
after verse, Muslims are directed to provide for those who cannot provide for themselves:

> *By no means shall you attain to righteousness until you spend benevo-*
> *lently out of why you love; and whatever thing you spend, Allah surely*
> *knows it. . . .*
>
> *And they ask you as to what they should spend. Say: what you can*
> *spare. . . .*
>
> *Who amasses wealth and considers it a provision against mishap;*
> *He thinks that his wealth will make him immortal. Nay! He shall*
> *most certainly be hurled into the crushing disaster. . . .*
>
> *And keep up prayer and pay the poor-rate (charity) and whatever*
> *good you send before for yourselves, you shall find it with Allah; surely*
> *Allah sees what you do. . . .*

Surely they who believe and do good deeds and keep up prayer and pay charity they shall have their reward from their Lord, and they shall have no fear, nor shall they grieve.

Zakat (charity) is one of the Five Pillars of Islam. One of the most important principles of Islam is that the earth and everything on it belongs to Almighty God. Humans hold it in trust. Nothing we have is truly ours individually.

There is no ambiguity about the requirement of Muslims to be charitable. There should be no ambiguity about helping to improve the quality of life of fellow Muslims. Since poverty breeds extremism, militancy, and terrorism, it is in the common interest of the Muslim nations to provide some sense of equalization amongst the respective economies of the Organization of the Islamic Conference. Al Qaeda is as much a threat to the Kingdom of Saudi Arabia as it is to the United States of America. The oil-producing states of the Muslim world could organize a fund, investing the income of the fund to stimulate economic development in the non-oil-producing Muslim member countries. Significant progress could be made in developing those economies. Actively developing economies would stimulate the emergence of a middle class, reversing the cycle of poverty that hurts generations in the economically poorer but demographically larger countries of the Muslim world. Here again, Pakistan, Afghanistan, Bangladesh, and Indonesia are good examples of target nations.

Thus I propose the creation of a Muslim Investment Fund, structured in a manner similar to the Alaskan fund or the Norway fund. This fund, with contributions from oil-producing Muslim states, could quickly accumulate an enormous amount of money. Assuming that the major oil-producing states in the Muslim world would sign on and set aside half the percentage of money that the Alaska Permanent Fund takes from its oil production, a Muslim Investment Fund would produce nearly $12 billion in the first year alone. If oil production levels remain the same in the medium term, in just twenty years the capital of the Muslim Investment Fund would be approximately $465 billion!

Once again, using the Alaska Permanent Fund as an example, let us assume that the Muslim Investment Fund would achieve a similar interest return per year from its various investments: 11 percent. These investments could serve two purposes. First, they would help the fund grow. Second, the money could be invested in the Muslim world, spurring economic growth. This assumed 11 percent return would spin off $1.25 billion the first year, and in year twenty, the Muslim fund would have produced tens of billions of dollars in interest. This income would belong to the donors to the Muslim fund. They could use it partially to offset poverty in the Muslim countries. Every year, as the capital endowment of the fund grew, the amount of interest earned from the fund would grow. Initially the interest earned could be reinvested into the fund to help the capital endowment grow. However, as with the Alaska fund, after a number of years, the yearly interest, an enormous sum, could be invested in Muslim countries to speed economic development and quality-of-life issues, while the capital endowment would never be touched. For those living below the poverty level throughout the Muslim world, the Muslim fund could mean the difference between a comfortable life and continuing poverty. And since—once again—we know that the hopelessness of the cycle of poverty breeds extremism and fanaticism, breaking the cycle would douse the fire of the militants. The Muslim fund could also create an intellectual renaissance for the Muslim world. It is notable that the fifty-seven member countries of the Organization of the Islamic Conference have approximately 500 universities, compared to 5,000 universities in the United States and 8,000 universities in India. In a compilation of the academic ranking of world universities conducted in 2004 by Shanghai's Jig Tong University, not a single university from the Muslim world was included in the top 500 universities on earth. In addition, the Muslim world spends 0.2 percent of its GDP on research and development, while the Western nations spend more than 5 percent of their respective GDPs on producing knowledge, generating ideas, and creating innovation.

The Muslim world's decline is not due simply to the injustices of colonialism or the global distribution of power. At some point Muslim societies must be responsible and accountable. There is an abundance of riches in Muslim countries. If organized properly, the Muslim countries could draw up an agenda to reduce poverty and rekindle Islamic nations as centers of knowledge and ideas. The Muslim countries have the power to change the direction of history by adhering to the Islamic teaching of sharing wealth. Such collective action is a challenge that we must have the courage to face and act upon. By creating a larger national Muslim response through a Muslim fund, our generation can raise revenues for the benefit of generations of Muslim children as yet unborn.

·

The clash within Islam is but a part of the problem threatening world stability. There is another potential element of international disruption that some might call "the global war on terror" and others hyperventilate into "World War IV." Even if the Islamic nations were to do everything that I suggest, we would still be left with a significant chasm between the Islamic countries and the West. There is a strong sense in the Muslim world that the West wishes to impose its values on Muslim societies and that these values are often inappropriate or decadent and undermine Islamic values of family life. It doesn't matter if this is true or not. If it's believed, it becomes self-fulfilling because perception often shapes reality.

The question before the West is twofold. First, the West should look inside and determine to what extent Muslims' perceptions of the West are justified, or at least understandable. Just as I urge my fellow Muslims not to blame others for problems that we are at least partially responsible for, I urge the West not to blame Muslims for problems that have arisen partially from the West's culpability. And second, the West must open up in considering what steps can be taken to bridge the chasm between societies and cultures.

The West must be ready to acknowledge the residual damage of colonialism and its support for dictatorships during the Cold War. The West, and especially the United States of America, must be ready to revisit the rippling impact of the so-called global war on terror, which is perceived by perhaps hundreds of millions of Muslims as a "global war on Islam." Many Muslims fear that the Islamic faith is under attack and Muslims are under siege. They believe that the conflict in Iraq is part of a much broader (and nefarious) Western agenda. Before the problem can be solved, we need to recognize that it exists.

Introspection is never easy and almost always uncomfortable. But in the current international environment, a period of introspection by the West is necessary. It is critical for the West—and, most important, the United States—to examine the extent to which Islamic concerns and criticisms are justified and then commit to addressing these concerns substantively. I am not condoning terrorism or hatred. But a problem existed before September 11, 2001, and that problem will continue to exist after Al Qaeda is but a painful memory. There is confusion between the West and the Islamic world, and there is, to some extent, distrust. The confusion can be clarified. The distrust can be overcome. But for that a plan is needed—and action.

.

The lessons of history help us plan for the future. The conditions, threats, and opportunities that confronted Europe at the end of World War II can give us guidance on how to intelligently and effectively address the current situation we find ourselves in with respect to Islam and the West. For this, I turn to the words of U.S. Secretary of State General George Marshall, delivered at my alma mater, Harvard University, on June 5, 1947. He could have been talking about our current crisis. His solutions proposed at Harvard are relevant, in many ways, to our present times.

Marshall noted:

The world situation is very serious. That must be apparent to all intelligent people. I think one difficulty is that the problem is one of such enormous complexity that the very mass of facts presented to the public makes it exceedingly difficult for people in the street to reach a clear appraisement of the situation. Furthermore, the people of the country are distant from the troubled areas of the earth and it is hard for them to comprehend the plight and consequent reactions of the long-suffering peoples, and the effects of those reactions on their governments in connection with our efforts to promote peace in the world.

Marshall was speaking of Europe, but his comments are striking. He also said that "it is logical that the United States should do whatever it is able to do to assist in the return of normal economic health in the world, without which there can be no political stability and no assured peace. Our policy is directed not against any country or doctrine but against hunger, poverty, desperation, and chaos." Marshall specifically called for an understanding of the character of the problem and proposed remedies for application. Brilliantly (and strategically), he declared that "political passion and prejudice should have no part." From that commencement speech at Harvard emerged a $20 billion commitment by the United States to rebuild Europe and, in doing so, to preserve its own security. The Marshall Plan was both moral and self-serving, which is the key to defining national interest. That same formula could be applied to the Muslim world by the countries of North America, Europe, Australia, China, and Japan. This would comprise a new commitment pledging to eliminate terrorism within Muslim nations by systematically attacking the economic, social, and political roots of extremism.

I propose a new program by the developed world similar to the Marshall Plan, specifically using tangible and identifiable means to improve the lives of people in deprived areas of the Muslim nations. I am looking for programs whose success can be measured and evaluated. When ordinary people in a country identify assistance improving their lives and the lives of their children, they bond with the source of that aid. Such a bonding could bring a dramatic turnaround in perception among Muslims—on the so-called Muslim street—about the West. Empirical evidence substantiates this. After the horrific tsunami devastated Indonesia in December 2005, American public and private sector aid poured into the most populous Islamic nation on Earth. A poll taken by Terror Free Tomorrow immediately thereafter showed that favorable views of the United States jumped 65 percent because of American aid to the Indonesian people. Programs help change perceptions.

A similar phenomenon was found in Pakistan following the 2005 earthquake that killed almost 90,000 of my countrymen and women. The United States alone committed half a billion dollars in relief, and American military transports and American soldiers delivered that assistance to freezing and starving survivors. Polling conducted by ACNielsen Pakistan showed that favorable opinion of the United States doubled in response to the visible, generous assistance. Correspondingly, the same polling indicated a precipitous drop in support for Osama bin Laden and Al Qaeda over the same period, explainable by the same phenomenon of direct and visible U.S. support.

These kinds of data have been used by scholars to question whether the level and depth of anti-Western sentiment in the Muslim world is sufficiently broad and deep that it really will take generations to undue, as some say. It appears that support—of the right kind and in the right way—generates a dramatic sea change in attitudes quickly. This strengthens the argument for a different and expanded approach to development assistance by the West to the Muslim world. In the past, Western assistance was funneled largely through governments, mostly authoritarian. But contacts with Westerners on humanitarian missions substantially alter Muslim public attitudes. Direct engagement helps beneficiaries fully understand who the benefactors are. Foreign assistance can alter public attitudes if a "middleman" is not there to cause confusion about the source of the aid.

The Marshall Plan's $20 billion commitment in 1947 would now be equivalent to $185 billion. It is a formidable sum of money. However, if the plan were to be shared by North America, the European Union, Japan, and China, the funding would become less prohibitive. Moreover, it is estimated that the United States has already spent $500 billion on the Iraq War without improving the image of the United States or the West abroad, especially in the Muslim community. The total costs of the war—including care for injured soldiers for the rest of their lives and a continued U.S. presence in Iraq for the foreseeable future—could total $2 trillion when all is said and done. A Marshall Plan level of commitment of $185 billion in 2007 dollars pales by comparison.

I am not proposing a program of writing checks to governments. I am proposing specific and tangible people-to-people projects that will directly improve the quality of life of ordinary people, in the form of humanitarian aid from the West. I recall that as prime minister I was able to accomplish much good by the personalization of the antipolio campaign that I introduced in my country. I was absolutely shocked to learn that Pakistan and Afghanistan together accounted for three-quarters of new polio cases in the world in 1993. I determined to do something about it. I administered the antipolio drops to my daughter Aseefa. I invited Pakistani mothers with children born at the same time as Aseefa to join me at the prime minister's house to administer their children's drops. The program spread across the country with great fanfare, into every town and village. I am very proud that that program helped eradicate polio in Pakistan. There were no new cases of polio in my country last year, and this success is the result of a specific, tangible program that I initiated. This is the model I propose for a twenty-first-century Marshall Plan to assist the Islamic world to leap into modernity.

The personalization of direct and specific programs can and will have tangible results in spreading goodwill and influence. In the negative, I specifically recall that the Al Rasheed terrorist network, later found to have been involved in the kidnapping and murder of the journalist Daniel Pearl, set up tandoor oven bakeries throughout the Taliban-controlled areas of Afghanistan. Each day mothers and fathers would come to the bakery to pick up their families' daily allotments of nan, three pieces for every family member. These families had been starving, and now they were being fed. And in every tandoor bakery set up by Al Rasheed in Afghanistan, a picture of Osama bin Laden looked down from the wall. As parents got their rations of bread that would keep their children alive, it was Osama bin Laden who got the credit. Thus goodwill toward the Taliban and Al Qaeda rapidly expanded through Afghanistan. Once again the Taliban controls a large part of the territory of Afghanistan. In part it is due to the goodwill generated by the delivery of social service programs to the poor, hungry, ill clothed, and ill housed. If this model is replicated throughout the Muslim world—with the international community delivering services and receiving the credit from the people—much goodwill can be generated to help Islamic-Western relations in the years ahead.

I am not proposing huge and complex programs. I am proposing simple, clear, visible, and pervasive programs, giving people what they absolutely need for their family that does not require middlemen who skim product and skim credit. I'm proposing grain, schoolbooks, medicines, writing materials, and inexpensive shoes. I am proposing the same level of immunization against other diseases that we so successfully accomplished against polio, with each little bottle of serum clearly marked with the country of origin. Of course building modern hospitals is welcome, but another place to start is training health-care workers to go out into the villages one at a time to provide basic health services and supplies to rural villages that have no doctors, much less a hospital. I did it when I was prime minister with women health workers who gave information on family issues, preventing infant mortality, and family planning, and it was remarkably successful.

I want those kinds of manageable but widespread and visible programs expanded. I would recommend programs that deliver clean drinking water, provide rudimentary public housing, and build one-room rural schools. These simple, rudimentary programs not only are necessary but would be greeted with gratitude and joy by hundreds of millions of deprived Muslims from Bangladesh to Malawi. And to the extent that the international directors who administer these programs on the ground can speak the language of the people—whether Arabic, Urdu, Bengali, or Indonesian Malay—the better for understanding the recipients. (It would be quite productive for the West to do far more in language training for its secondary school and university students, as well as its Foreign Service officers, in the various languages of Muslim-majority countries.)

North Americans, Europeans, Chinese, and Japanese should view this program to help fellow human beings not as charity but rather as an investment. This should not be viewed as acquiescence to the threat of terrorism but the kind of wise pragmatism that enabled America to rebuild its European World War II allies, as well as its former Axis adversaries of Germany, Japan, and Italy, in the late 1940s. The Marshall Plan was created and implemented because it made sense for global peace. It was calculated and methodological. And in the words of George Marshall himself, "Political passion and prejudice should have no part." The Marshall Plan addressed not only what was good but what was smart. It spoke to those who wanted the United States' foreign policy to be based on morality, as well as those who believed that policy should be driven by realpolitik. It provided food and shelter to the hungry in Europe while containing communism. An International Fund for Muslim Economic and Social Development could achieve positive moral objectives and pragmatic political goals for a more peaceful world.

·

Economic reconstruction can help to turn the Muslim street around. However, another fundamental way to contain the spread of extremism and militancy is the reinforcement of democracy around the world. Inspired by George Marshall's unveiling of the Marshall Plan at the Harvard commencement of 1947, I used the great honor of my commencement speech to the Harvard graduating class of 1989 to propose another bold foreign policy initiative that I believe is vital to international stability and peace. I proposed the creation of an Association of Democratic Nations to band together to protect and strengthen young democracies around the world. After I left office, President Corazon Aquino of the Philippines headed an organization based on the concept called the Community of Democratic Nations, but it never really took off. We need either to empower the CDN or to start a new organization that has the power and influence necessary to help sustain fragile and vulnerable democracies. I think it is more important than ever to rekindle and reenergize that concept, that program, that goal in the political environment we face today. One election does not make a democracy. Elections can lead to the rise of fascist leaders who persecute the opposition, intimidate the press, and crush the judiciary. I saw it in Pakistan. Civilian and military dictatorships are both bad.

Democracy needs support. The best support for democracy can come from other democracies. The annual report on human rights of the U.S. State Department reports on these conditions in countries around the world. As a political prisoner of the military dictator of Pakistan in the 1980s, I was heartened to learn that the U.S. Congress had linked assistance to Pakistan to "the restoration of full civil liberties and representative government in Pakistan." Friends of democracy in other countries, including Britain, Canada, and Germany, do send delegations to investigate human rights abuses in Pakistan. Election monitors from democracy-monitoring organizations from the United States, the British Parliament, and the South Asian Association for Regional Cooperation do visit young democracies. This informal network of democratic NGOs can make a critical difference in many other nations struggling to restore or adopt democracy. This informal network can be institutionalized and strengthened. Democratic nations should forge a consensus around the most powerful political idea in the world today: the right of people to freely choose their government and for governments so selected to govern democratically pursuant to the rule of law.

The informal bond should translate into a formal organization designed for democracies to help other democracies sustain democracy. Such an Association of Democratic Nations could provide governmental support for the international civil society's efforts to protect and defend democratic elections and monitor democratic governance. One way the association could be vital is ensuring the impartiality of elections—it could become an international analogue to the National Endowment for Democracy, using the skills and resources of human rights and political rights training and monitoring groups around the world to observe and legitimize or delegitimize elections around the world. For a verdict to be accepted as legitimate, an election must not only be fair, but it must also be seen to be fair. The presence of observers is a deterrent to fraud. Observers bring television cameras with them. It is harder to steal an election if the whole world is watching. That certainly was the lesson of the Philippine election of 1986, which ultimately resulted in the certification of Corazon Aquino over the Ferdinand Marcos regime. Attempted fraud under the glare of television lights can help galvanize a popular uprising.

There are other ways in which such an international Association of Democratic Nations could provide some protection for democratic governments. In countries without established traditions of representative government, democracy is always at risk. All too often, there is an overly ambitious general, a determined fanatic, or an ambitious civil leader who wants to become a civilian dictator. They are threats all over the world but a particular threat to nascent and fragile democratic experiments in the Muslim world. The Association of Democratic Nations could help change the calculus for potential coup plotters by adding the element of international opprobrium. Perception does matter, even to tyrants.

The association's power would flow from its ability to delegitimize fraudulent elections and isolate fraudulent governments. Public persuasion can be powerful, but often it isn't enough. When General

Musharraf declared martial law in November 2007 and the entire democratic world denounced his actions, he didn't seem to particularly care. When Pakistan was suspended from the Commonwealth, and when the U.S. Congress signaled that it would consider conditioning military aid to Pakistan on the country's transition back to democracy, General Musharraf and his men took notice.

My proposal seeks some order to the process of political legitimization. Presently, the pattern of election monitoring is erratic. The Commonwealth may or may not be involved. SAARC is often involved but rarely criticizes the actions of its members. The Organization of Islamic Countries (OIC) is not yet credible on election monitoring. American political institutes are helpful but have limited resources. The international community of democratic nations needs to assert itself as democratic nations. The United Nations is constrained by the universality of its membership. When violators of human rights can sit on U.N. human rights commissions, the work of such bodies has no authority or legitimacy. An Association of Democratic Nations would be free from such problems. It could objectively evaluate political systems. It would be especially welcome in the large-population Muslim countries of Pakistan, Bangladesh, and Indonesia, where dictatorship-versus-democracy problems are constantly raised.

·

A similar bringing together of international organizations could be a powerful tool for protecting the institutions of civil society in transitional democracies, especially in Muslim-majority nations. Although individual elements of civil society are well organized around the world, they lack a mechanism to draw their influence together for common purposes. An aggregation of civil society organizations would exercise enormous power and influence, especially in protecting fragile civil society institutions in new and emerging democracies.

International aggregations of journalists, human rights organizations, judges, lawyers, bureaucrats, academics, students, doctors, and scores of professional groups exist that could be mobilized under a powerful umbrella. Human rights groups such as Amnesty International, Human Rights Watch, UNIFEM, the International League for Human Rights, Freedom House, Physicians for Human Rights, the International Human Rights Association, the Carter Center, the International Helsinki Commission for Human Rights, the World Organization Against Torture, Reporters Without Borders, the International Association of Democratic Lawyers, the Committee to Protect Journalists, Doctors Without Borders, the International Committee of the Red Cross, the World Health Organization, Project Hope, Health Volunteers Oversees, and the Global Network of People Living with HIV and AIDS could be bonded into a civil union that would inspire enormous respect and influence. Development groups such as Oxfam International, Africare, the World Bank, the International Monetary Fund, the Association for International Agriculture and Rural Development, and the International Association of Science and Technology for Development could also be brought into this powerful tent.

Together these international civil society groups could serve as the foundation of a civil society analogue to the United Nations—a Civil Society International (CSI). This new Civil Society International could respond to attacks on NGOs around the world and bring the full weight of international pressure to persuade offending governments to rescind their orders and restore the rule of law as it pertains to the civil society.

Such a Civil Society International could have acted immediately after November 3, 2007, when General Musharraf declared martial law in Pakistan, suspended the Constitution, arrested Supreme Court justices as well as thousands of lawyers and other judges around the nation, imprisoned hundreds of human rights and women's rights activists as well as political party leaders, and shut down independent electronic media across the country.

·

The perception of the West by the Muslims of the world is deeply problematic. The problem is particularly intense concerning their feelings toward the United States. An abundance of data shows a steady and dramatic deterioration of approval of the United States in Muslim societies spanning at least five years. The Pew Research Center has done consistent work on this topic demonstrating how formidable the problem of reconciling Islam and the West will be, in particular Muslims and the United States.

The data demonstrate that America's public relations problem is not limited to Muslims. From 2000 to 2006, respect for the United States even among its closest allies dropped precipitously. A few examples are telling. Favorable opinion of the United States over the six-year period dropped from 83 to 56 percent in Great Britain, from 62 to 39 percent in France, from 78 to 37 percent in Germany, and from 50 to 23 percent in Spain. Of course, in the Muslim world that Pew sampled, the negative perception of the United States is considerably worse. Favorable opinion of the United States is now 30 percent in Indonesia and Egypt, 27 percent in Pakistan, 15 percent in Jordan, and an extraordinarily low 12 percent in NATO member Turkey. In these same countries, concern that the United States is a military threat to the respective countries ranges from 80 percent among Indonesians, who are "very worried" about a possible American attack, to 67 percent of Jordanians. The result was characterized by Pew's Andrew Kohut before the U.S. Congress quite dramatically: "After Iraq, many people in Muslim countries began to see the United States as a threat to Islam, and what had perhaps been simple loathing for the United States turned into both fear and loathing."

The more detailed data are equally disturbing. The U.S.-led war on terror is perceived in much of the Muslim world as a war on Islam. More than half of Moroccans and half of Jordanians believe that suicide attacks against Westerners are justifiable. The long-range problem seems to center on the restoration of trust. According to Kohut, "the challenge is how to reverse the impact of images of Abu Ghraib and Guantanamo that now shape the views of young people all around the world, as favorable depictions of America as defender of freedom in the 20th century did then." Well, I have some thoughts about that.

The Djerejian Report "Changing Minds, Winning Peace," issued in 2003 and funded by the U.S. Congress, acknowledged that one great flaw in U.S. public diplomacy was the failure of the United States to carefully listen to the target audience in the Islamic world and to engage in genuine dialogue. American public diplomacy failed to adjust to the new communications age, where people are bombarded by information twenty-four hours a day, seven days a week, on multiple, overlapping mechanisms: public television, private television, cable television, print, radio, and the Internet. In other words, information cannot be controlled to preach a single political ideology. The United States and the West have been out-performed and out-innovated by extremists preaching hatred and violence.

James Traub made a similar critique of U.S. public diplomacy in *The New York Times.* He observed that "the weapons of advocacy had fallen into a long decline since their heyday in the cold war, a contest that pitted political systems against each other more than armies." He argued that the political weapons of the Cold War are irrelevant and inappropriate to the current "battle of ideas" in the world, especially the intellectual assault of Muslim extremists. What Traub and others recognize, and what some Western governments do not, is that what a country says is no longer as important as what it does. It is policy that ultimately determines respect and influence. The West has a great product: democracy, diversity, pluralism, tolerance, intellectual debate, accountability, a methodology for peaceful governmental change. This was always the power of the West, the power of "we, the people." The attractive features of open, free Western societies cannot be sold like a box of cereal; they cannot be packaged in glossy brochures. The West's attractiveness rests in exposing people—elite opinion leaders and millions of young people of all social classes—to the richness of the Western political experience, the generosity of its people, and the ability to allow those with talent to rise to leadership positions irrespective of their race, religion, or creed.

The Educational and Cultural Exchange Program of the U.S. Department of State has brought tens of thousands of young foreigners, many destined to become leaders of their nations, to the United States to study, to visit, to learn. Invariably, those exposed to these exchanges become great advocates of free and accountable societies when they return home. Those who have become presidents and prime ministers, like myself, have applied the lessons of democracy, gender equality, and freedom of speech learned in study to their own nations' processes of democracy—both in elections and in governance. And our warmth toward our fellow students lives on to this day.

It is quite remarkable that when the former Soviet Union tried to emulate the U.S. education and exchange model, the programs boomeranged. Third-world students brought to Patrice Lumumba University in Moscow typically returned home hating communism and the Soviet system. What ultimately "sells" is the product, and the West's greatest "product" is not its technology, not its weapons systems, not even its clothes and music, which are emulated all around the world. What ultimately makes Western capitals the "shining cit[ies] on the hill" to billions of people around the planet are their political and social values of tolerance and freedom. And that is what should be the key to public diplomacy today.

The exchange programs that have been effective should be expanded and strengthened. The West should be committing significantly more resources than it now commits to dialogue and intercultural understanding. It needs to be using the Internet creatively and in ways that young people will access and relate to. After all, "Google" is an American verb.

The mechanisms of public diplomacy must also be adjusted to the new environment. Scholarships to bring Muslim students to Western colleges and universities to learn in an open intellectual environment while getting first-rate educations in their chosen fields should be greatly expanded. Youth exchange programs to bring high school students to the West to participate in summer programs contribute to a more peaceful world.

International visitors programs to bring young leaders for short-term study tours and homestays to witness Western life firsthand while developing global professional networks is another tool of understanding that can be dramatically expanded. The same phenomenon results from the citizen exchange programs, which build expertise in civil society institutions. The Humphrey Scholarship Programs, which bring midlevel bureaucrats to work in American state and national government, help people from different parts of the world to "network" within their professions.

English-teaching programs still exist but need extensive expansion. They provide training and material resources to promote English teaching abroad, to foster professional skills and job opportunities. The British Council does an extraordinary job in this respect, and its success should be emulated by others.

Exhibits that once attracted tens of thousands of visitors to embassies and cultural centers have now been eliminated. They generated extraordinary goodwill, and it's a pity that they were done away with. They allowed people across continents to appreciate the universality of social values. Such exhibits should be restored.

The West should help provide computer equipment and computer training to students throughout the Muslim world. Exposing young Muslims to a flood of information would open their minds and hearts. It would be politically invaluable. Under the recent martial law in Pakistan, television and radio were shut down and the printed media censored, but the Internet kept the flow of information going. It kept people aware of events as they unfolded.

Providing rural schools in Pakistan with computer equipment and training would be a dramatic step forward to exposing young people to a new world. The more they know, the more they learn, the more they will challenge the status quo. And their challenging of the status quo will echo the words of George Bernard Shaw that my friend Kathleen Kennedy's father, the late Robert F. Kennedy, so powerfully quoted across America in 1968: "There are those who look at things the way they are, and ask why. . . . I dream of things that never were and ask 'why not?'" Young Muslims all over the world want to start asking—and dreaming and demanding opportunity and change. Access to information ended autocracy in Eastern and Central Europe. It can open up societies in the Muslim world as well.

.

Another innovation that could help to restore communication, trust, and dialogue between the Muslim world and the West would be the creation of a Reconciliation Corps. Modeled on the Peace Corps, it would be made up of Muslims from Western societies who have been economically, socially, and politically integrated into the life of their host countries while maintaining their Islamic character, culture, and religion. These Muslim youths could build bridges with their countries of origin.

The standard of living of Muslims in the West is extraordinarily high. In many communities it is equal to or above that of the non-Muslim populations. Universities and schools of higher education have high numbers of Muslim students, and the income level of Muslims in the United States, Canada, and Australia is parallel to that of the non-Muslim communities there. Not only have Muslims felt generally welcome in these Western countries, but there can be little dispute that they have been completely free to practice their religion and culture. Indeed, Muslims in the West are certainly freer to practice their religion than are non-Muslims in Muslim countries. If we are looking for models of reconciliation that prove that Muslims and Westerners can not only coexist together but also thrive together, we should look to the United States, Canada, and Australia.

A Reconciliation Corps of young Muslim professionals from Western nations could be trained as teachers, health-care practitioners, and computer technicians. They could be posted in Muslim-majority countries to provide essential services to the local populations with whom they would live. They could thus demonstrate to fellow Muslim populations that Muslims, Christians, Jews, and Hindus can successfully live together in peace. The Reconciliation Corps could be an agent of perceptional change, breaking down barriers and building new bridges.

As a product of a religious partition in 1947, I have always been struck by the similarities between the situation on the South Asian Subcontinent and in Palestine. Each Partition created two nations—Palestine and Israel, and India and Pakistan. The Partition was accepted on the Subcontinent but rejected by one of the parties in the Middle East. Whether that rejection was wise or not is irrelevant. What is relevant is that today the two-state solution is accepted by all reasonable parties to the dispute, marginalizing extremists on both sides. A serious effort to resolve the Palestinian issue would be a significant step toward, at the very least, eliminating one hot-button issue that is often used to put other very significant issues on the back burner.

It is clear that a Palestinian state will eventually be established in Gaza and the West Bank, and I believe the sooner, the better. Territorial and geographical issues can and will be worked out. One very significant issue that may prove most difficult is the resolution of the grievances of Palestinian refugees forced from their homes in 1948. The Arab states, meeting in a summit conference in Beirut in 2002, accepted the Riyadh Declaration, which acknowledged the existence of the state of Israel conditioned on something that Israel appears unlikely to accept: the return of 2 million Palestinian refugees to Israeli territory. One hopes that having come so far, the Arabs and Israelis will be able to find a solution to this issue, too.

But Palestine is not the only issue of concern to Muslims that has been allowed to fester over the years. Kashmir is a problem that needs to be solved, especially since it is an emotional conflict between two nuclear-armed neighbors that have been adversaries in the past. We cannot go back sixty years to whether the people of Kashmir, pursuant to the terms of Partition and to the mandate of U.N. resolutions, should have been allowed to determine their own future. They were denied that right, and for the last sixty years the state of Jammu and Kashmir has been a catalyst for war, anger, and militancy.

India and Pakistan, with the backing of the international community, have engaged in quiet diplomacy for several years in what is known as the "composite dialogue." A resolution to this conflict would remove one more source of Muslim discontent and anger while ending cross-border military actions that always threaten to degenerate into a ground war or worse.

Finally, the situation in Chechnya should be maturely addressed by the Russians, hopefully with the participation of the OIS and the United Nations. Chechnyan Muslims feel occupied and brutalized, not unlike how Palestinians in the West Bank and Gaza felt before the creation of the Palestinian Authority. There is no reason to think these passions will disappear. The reasons for the passion need to be addressed in a way that gives Chechnyans a sense of autonomy in a manner that is geographically and politically acceptable to Russia. This too can be done, but it requires the commitment of the parties working in a spirit of goodwill. If it is resolved, it would remove the third major rallying point for those who work for "clash" instead of for peace.

·

In these writings I have tried to trace the roots, causes, and potential solutions to the crisis within the Muslim world and the crisis between the Muslim world and the West. Theology, history, economics, democracy, and dictatorship have all played significant roles in bringing the world to this crossroads. My premise from the beginning has been that extremism thrives under dictatorship and is fueled by poverty, ignorance, and hopelessness. The extremist threat within the Islamic world and between the Islamic world and the West can be solved, but it will require addressing all the factors that breed it.

I appreciate that what I propose—from what the Muslim states must do to what the West must do—is huge and may seem daunting and even impossible. I make these recommendations because the times require something more than business as usual. Much of what is recommended is somewhat out of the box. But staying within the box has brought poverty, ignorance, hopelessness, violence, and dictatorship to far too many Muslims around the world. Staying within the box has set Islam and the West on a dangerous and unnecessary collision course. It is time for new ideas. It is time for creativity. It is time for bold commitment. And it is time for honesty, both among people and between people. That is what I have tried to do in these pages. There has been enough pain. It is time for reconciliation.

Afterword

There was a reason that Allah gave our wife and mother the time to finish this book.

This book is about everything that those who killed her could never understand: democracy, tolerance, rationality, hope, and, above all, the true message of Islam. Or maybe they did understand these things and feared them, and thus feared her. She was the fanatics' worst nightmare.

Nothing can lessen our grief or mend our broken hearts. The tears, loss, loneliness, and longing to be with her, hear her voice, and receive her guidance, all of this will be part of our souls forever.

Those who loved her and will always love her must keep fighting for the things in which she so strongly believed and for which she was willing to risk her life. We will continue the battle for democracy and against extremism and hatred. We commit our lives to making the message of this book into her legacy and the future of a democratic Pakistan. And ultimately we know we will succeed because, in her own words, "Time, justice, and the forces of history are on our side."

Jeay Bhutto. Bhutto lives.

ASIF ALI ZARDARI
BILAWAL BHUTTO ZARDARI
BAKHTAWAR BHUTTO ZARDARI
ASEEFA BHUTTO ZARDARI
Naudero, Pakistan
January 3, 2008

Acknowledgments

There were many people who assisted in this project to whom we are extremely grateful:

To Asif, Bilawal, Bakhtawar, and Aseefa Zardari, with the deepest love and thanks for everything—large and small—and for the irreplaceable gift of precious time.

To Nusrat Bhutto and Sanam Bhutto, loving mother and sister, with gratitude and love.

To Judith Siegel, for her support and encouragement, for asking the tough questions, and for editing draft after draft in a true labor of love to both the author and collaborator.

To Husain Haqqani, who played a vital role in providing guidance and criticism, especially invaluable on the theocratic foundations of Islam and the history of Pakistan—a loyal friend whose counsel to both the author and the collaborator will always be cherished.

To our research assistant, David Knoll, who worked tirelessly and efficiently to address our ever-expanding needs and requests and made an important contribution to this book.

To Brigadier Amanullah (Benazir Bhutto's chief of staff in Islambad), who served for eleven years after his retirement out of his commitment to the future of his country.

To Musarrat Alil at Zardari House in Dubai and Irfan Ali Larik at

Bilawal House in Karachi for everything they have done over the years, and for working on incorporating edits to this book.

To Roy Coffee, Phil Rivers, Dave Distefano, Shane Doucet, and Victoria Collins for their support and patience. And very special thanks to Christa Natoli. To Carl Gershman and the staff of the National Endowment for Democracy, for their suggestions and counsel.

To Faratullah Babar, Sherry Rehman, and Farahnaz Ispahani for their power of communication.

To John Dougherty, our research intern, for his hard work.

To Wendy Chamberlin and the Middle East Institute, for logistical support, cooperation, and sponsorship of Benazir Bhutto's speech in the Russell Building of the United States Senate on September 25, 2007.

To Christopher Arterton and the Graduate School of Political Management of the George Washington University, for their research and logistical support.

To our dear friend Peter Galbraith, whose introduction of two people twenty-four years ago made this book possible.

To Victoria Schofield, an absolute rock of friendship, who was on the truck in the caravan on the momentous day of October 18, sharing as always both the triumph and the tragedy.

To Andrew Wylie, agent supreme, but more important a loving friend, whose encouragement and persistence convinced us to take on this formidable project against all odds.

To Tim Duggan, for his skillful, patient, and wise editing. This project was conducted under the most unusual and difficult conditions, and his guidance made all the difference.

·

And finally, a special word of tribute to all of those who gave their lives to the cause of democracy in Pakistan in Karachi on October 19, 2007, and in Rawalpindi on December 27, 2007: you will not be forgotten, and you, in the end, will prevail.

Notes

1. THE PATH BACK

3 "think it is good": Andrew Kohut, "America's Image in the World: Findings from the Pew Global Attitudes Project," testimony before the Subcommittee on International Organizations, Human Rights and Oversight, Committee on Foreign Affairs, United States House of Representatives, March 14, 2007, available at www.pewglobal.org/commentary/display.php?AnalysisID=1019.

9 "Have you noticed": Christina Lamb, "It Was What We Feared, but Dared Not to Happen," *Times* (London), October 21, 2007.

2. THE BATTLE WITHIN ISLAM: DEMOCRACY VERSUS DICTATORSHIP, MODERATION VERSUS EXTREMISM

18 "mutual advice through": Fazlur Rahman, "The Principle of Shura and the Role of the Ummah in Islam," in *State, Politics, and Islam,* ed. Mumtaz Ahmad (Indianapolis, Ind.: American Trust Publications, 1406/1986), 90–91.

19 "going astray from the": Quran 5:77.

21 "The simplistic translation": Asma Afsaruddin, "Competing Perspectives on Jihad: Jihad and 'Martyrdom' in Early Islamic Sources," in *Witness to Faith?,* ed. Brian Wicker (Burlington, Vt.: Ashgate Publishing Company, 2006), 16.

21 Clearly there are some: John Esposito, *Unholy War* (New York: Oxford University Press, 2002), 27.

22 "We return from": Ibid., 28.

23 "And the recompense": Quran 40:40–41.

23 "And whoever is": Quran 40:43.

23 "Permission (to fight) is": Quran 22:39.

23 "And fight in the way": Quran 2:190.

23 "So when the sacred": Quran 9:5.

24 "Fight those who": Quran 9:29.

24 "And if they incline": Quran 8:61.

25 "In both Islam": Majid M. Khadduri, *War and Peace in the Law of Islam* (Baltimore, Md.: Johns Hopkins University Press, 1955), 57.

25 "Islam outlawed all": Ibid., 62.

25 "*Jihad* is a collective": Ibid., 60–61.

26 "When one of you": The Prophet quoted in Afsaruddin,"Competing Perspectives on Jihad: Jihad and 'Martyrdom' in Early Islamic Sources," 26.

26 "For this reason": Quran 5:32.

26 "He who disbelieves": Quran 16:106.

27 "And spend in the way": Quran 2:195.

27 "O you who believe": Quran 4:29.

27 "those who unilaterally": Afsaruddin, "Competing Perspectives on Jihad: Jihad and 'Martyrdom' in Early Islamic Sources," 20.

27 "the terrorism we": Osama bin Laden, interview with ABC News reporter John Miller, May 1998.

28 "Muslim citizens thus": Esposito, *Unholy War,* 46.
28 "and so Maudoodi sees": Ibid., 55.
28 He used the term: Ibid., 56.
29 "We do not have to differentiate": Bin Laden, interview with ABC News reporter John Miller, May 1998.
29 "So they went on": Quran 18:74.
31 "And if your Lord had pleased He would": Quran 11:118.
31 "And if your Lord had pleased, surely all": Quran 10:99.
31 "You shall have your religion": Quran 109:6.
31 "Surely those who believe": Quran 2:62.
32 "strive with one another": Quran 5:48.
32 "Whatever Allah grants": Quran 35:2.
32 "Those who bear the power": Quran 40:7.
32 "Will they distribute the mercy": Quran 43:32.
32 "There is no compulsion": Quran 2:256.
33 "If Allah had pleased He would": Quran 5:48.
33 "O people! be careful": Quran 4:1.
34 "And the same did Ibrahim": Quran 2:132.
34 "Nay! were you witnesses": Quran 2:133.
34 "Surely Allah chose Adam": Quran 3:33.
34 "Say: O followers of the Book!": Quran 3:64.
35 "Say: We believe in Allah": Quran 3:65.
35 "Say: Surely, (as for) me": Quran 6:161.
35 "And mention Ibrahim": Quran 19:41.
35 "Surely We revealed the Taurat": Quran 5:44.
35 "Again, We gave the Book": Quran 6:154.
35 "And most certainly We gave Musa": Quran 2:87.
35 "Say: We believe in God": Quran 2:136.
36 "He has revealed to you": Quran 3:3.
36 "Surely We have revealed": Quran 4:163.
36 "And We sent after them": Quran 5:46.
36 "And when We made a covenant": Quran 33:7.
36 "You belong to your father": Christian Bible (John 8:44).
36 "Woe to you, teachers of the law": Christian Bible (Matthew 23:29–34).
37 "The word of the Lord spread": Christian Bible (Acts 13:49–50).
37 "Men of Israel": Christian Bible (Acts 3:12–15).
37 "For God so loved the world": Christian Bible (John 3:16–18).
37 "The enmity between us": Osama bin Laden, interview with ABC News reporter John Miller, May 1998.
38 "When we say 'innocent people'": Anjum Chaudri, interview with Stephen Sackur on *Hardtalk,* BBC, August 10, 2005.
38 "perfected": Ann Coulter, interview with Donny Deutsch on *The Big Idea,* CNBC, October 8, 2007.
40 "by the time the *shari'at*": Asghar Ali Engineer, *The Rights of Women in Islam* (New York: St. Martin's Press, 1992), 57.
41 "O people! be careful": Quran 4:1.
41 "And surely We have": Quran 17:70.
41 "They have rights": Quran 2:228.
41 "If you take out uncovered": Sheik Taj el-Din al-Hilali, sermon given September 2006, "Excerpts of al-Hilali's Speech," available at http://news.bbc.co.uk/2/hi/asia-pacific/6089008.stm.
42 "O Prophet! say to your wives": Quran 33:59.
43 "Say to the believing men": Quran 24:30–31.
43 "Surely the men who submit": Quran 33:35.
44 "men shall have the benefit": Quran 4:32.
47 "And they ask you a decision": Quran 4:127.
48 "And if they are pregnant": Quran 65:6.

51 "the most serious reason": Frederick Denny, *An Introduction to Islam* (Upper Saddle River, N.J.: Pearson Prentice Hall, 2005), 326.

52 "All faithful Shiites": S. A. Nigosian, *Islam: Its History, Teaching, and Practices* (Bloomington, Ind.: Indiana University Press, 2004), 49.

57 "The Ummah (Muslim community)": Abdul Hamid Abu Sulayman, "Culture, Science and Technology: How to Respond to Contemporary Challenges," *American Journal of Islamic Social Science* 19, no. 3 (2002): 79.

58 "transformed the early": Ibid., 82.

59 "Seek knowledge by": Jalal al-Din 'Abd al-Rahman ibn Abi Bakr al-Suyuti, al-Jami al-Saghir min Hadith al-Bashir al-Nadhir (Damascus: Maktabat al-Habuni, n.d.), 1:143.

59 "The concept of knowledge": Medhi Golshani, "Islam and the Sciences of Nature: Some Fundamental Questions," *Islamic Studies* 39, no. 4 (2000): 598.

59 "And one of His signs": Quran 30:22.

59 "Say: Travel in the Earth": Quran 29:20.

60 "And He taught Adam": Quran 2:31.

60 "And Allah has brought": Quran 16:78.

60 "in almost every": Tong Soon Lee, "Technology and the Production of Islamic Space: The Call to Prayer in Singapore," *Ethnomusicology* 43, no. 1 (Winter 1999): 86.

61 "The call to prayer": Ibid., 93.

62 "Neither Islam nor": Robin Wright, "Two Visions of Reformation," *Journal of Democracy* 7 (April 1996): 221.

63 "the character of Muslim": Abdou Filali-Ansary, "Muslims and Democracy," in *Islam and Democracy in the Middle East,* ed. Larry Diamond, Marc F. Plattner, and Daniel Brumberg (Baltimore, Md.: Johns Hopkins University Press, 2003), 193.

64 "Islam and democracy": Abdul Karim Soroush, quoted in Wright, "Two Visions of Reformation," 224.

65 "Sharia is something": Ibid., 226.

65 "The text is": Ibid., 225.

65 "Islamic authority is": Laith Kubba, "Faith and Modernity," in *Islam and Democracy in the Middle East,* ed. Larry Diamond, Marc F. Plattner, and Daniel Brumberg (Baltimore, Md.: Johns Hopkins University Press, 2003), 264.

65 "scripture has influenced": Ibid., 265.

66 "the Quran remains": Abdulaziz Sachedina, *The Islamic Roots of Democratic Pluralism* (New York: Oxford University Press, 2001), 46.

67 "And (as for) the man": Quran 5:38.

67 "Under this conception": El Fadl, "Islam and the Challenge of Democracy," 36.

68 "the doors of *ijtihad*": Radwan A. Masmoudi, "The Silenced Majority," in *Islam and Democracy in the Middle East,* ed. Larry Diamond, Marc F. Plattner, and Daniel Brumberg (Baltimore, Md.: Johns Hopkins University Press, 2003), 259.

68 "God has revealed": Ahmad, Kurshid quoted in John L. Esposito and John O. Voll, *Islam and Democracy* (New York: Oxford University Press, 1996), 29.

70 "The election of the chief": Abdul Rashid Moten, *Political Science: An Islamic Perspective* (New York: St. Martin's Press, 1996), 113–14.

70 "O you who believe!": Quran 4:59.

71 "Thus it is due to mercy": Quran 3:159.

71 "And those who respond": Quran 42:38.

71 "it signified, more broadly": El Fadl, "Islam and the Challenge of Democracy," 17.

71 "My community will": Esposito and Voll, *Islam and Democracy,* 28.

72 "Modern democratic opposition": Ibid., 39.

72 "And if your Lord": Quran 11:118.

72 "Surely Allah commands": Quran 4:58.

73 "Whatever Allah has restored": Quran 59:7.

73 "Our leader is one": Moten, *Political Science,* 116.

75 "Whereas sovereignty over": Preamble of the 1973 Constitution of Pakistan, available at www.helplinelaw.com/law/pakistan/constitution/constitution01.php.

3. ISLAM AND DEMOCRACY: HISTORY AND PRACTICE

83 "We will encourage": George W. Bush, Second Inaugural Speech, January 20, 2005, available at www.whitehouse.gov/news/releases/2005/01/20050120–1.html.

83 "a republic that supports": Noah Feldman, "Democratosis," *New York Times Magazine,* October 7, 2007, 11–12.

84 "Saudi Arabia—one": Feldman, "Democratosis," 11–12.

84 a relationship between: Carl Gershman, "Should the United States Try to Promote Democracy in the Middle East?," speech given on August 11, 2007, available at www.ned.org/about/carl/carl081107.html.9.

85 Each year it: Freedom House, "Freedom in the World," 2007 Edition, available at www.freedomhouse.org/template.cfm?page=15.

87 "almost 20 times": Alfred Stephan and Graeme B. Robertson, "An 'Arab' More than 'Muslim' Electoral Gap," *Journal of Democracy* 14, no. 3 (2003): 32–33.

87 "the largest single": Ibid., 30.

87 "no comparative Muslim": Ibid., 35.

109 "Saddam Hussein's Other": Mark Siegel, "Counter Point: Saddam Hussein's Other Republican Guards," *Wall Street Journal,* March 21, 1991.

109 The Western record: Peter Galbraith, *The End of Iraq: How American Incompetence Created a War Without End* (New York: Simon & Schuster, 2006).

131 attempts at inclusion: Gershman, "Should the United States Try to Promote Democracy in the Middle East?"

131 In Jordan, for example: Ibid.

139 "farce": "Presidential Elections in Tajikistan a Farce," Human Rights Watch 1999 Report: Tajikistan chapter, available at www.hrw.org/press/1999/oct/tajik1028.htm.

140 no true opposition: "Human Rights Developments: Uzbekistan," Human Rights Watch Report 2001, available at www.hrw.org/wr2k1/europe/uzbekistan.html.

154 "Communism was not": Václav Havel, "The End of the Modern Era," *New York Times,* March 1, 1992.

4. THE CASE OF PAKISTAN

168 "The defense of the State": Ian Talbot, *Pakistan: A Modern History* (London: Palgrave Macmillan, 2005), 118.

180 "a judicial inquiry": Talbot, *Pakistan: A Modern History,* 206.

184 Because of the qualified: Husain Haqqani, *Pakistan: Between Mosque and Military* (Washington, D.C.: Carnegie Endowment for International Peace, 2005), 97.

185 When Pakistan's surrender: Haqqani, *Pakistan: Between Mosque and Military,* 96.

188 When Zia overthrew: Haqqani, *Pakistan: Between Mosque and Military,* 112.

188 "often identified as": Haqqani, *Pakistan: Between Mosque and Military,* 131.

189 "supported military rule": Haqqani, *Pakistan: Between Mosque and Military,* 149.

192 "Zia left behind": Talbot, *Pakistan: A Modern History,* 286.

5. IS THE CLASH OF CIVILIZATIONS INEVITABLE?

234 multivolume grand review: Arnold Toynbee, *A Study of History,* vols. 1–6, abridgement by D. C. Somervell (New York: Oxford University Press, 1974).

235 "It should by now be clear": Bernard Lewis, "The Roots of Muslim Rage," *Atlantic,* September 1990, reprinted in *Policy* 17, no. 4 Summer 2001–2002: 26.

238 "the highest cultural": Samuel P. Huntington, "The Clash of Civilizations?" *Foreign Affairs* 72, no. 3 (1993): 24.

239 "a resident of Rome": Ibid.

239 "along the cultural": Ibid., 25.

239 "differences among civilizations are": Ibid.

239 "differences among civilizations have": Ibid.

239 "cultural characteristics and": Ibid., 27.

240 "even more so than ethnicity": Ibid.

240 "intensify civilization consciousness": Ibid., 25.

240 "longstanding local identities": Ibid., 26.

241   "dual role of": Ibid.
241   "return to the roots": Ibid.
241   "desire, the will": Ibid.
241   "Islam has bloody": Ibid., 35.
241   "this centuries-old": Ibid., 31–32.
241   "principal beneficiaries of": Ibid., 32.
242   "wherever one looks": Huntington, *The Clash of Civilizations and the Remaking of World Order,* 256.
242   "the West's simultaneous": Ibid., 211.
242   "so as long as": Ibid., 210.
243   "Kin Country Syndrome": Huntington, "The Clash of Civilizations?," 35.
243   "such conflicts . . . are": Ibid., 38.
243   "differences in culture": Ibid., 40.
244   "individualism, liberalism, constitutionalism": Ibid., 40.
244   "In Muslim eyes": Huntington, *The Clash of Civilizations and the Remaking of World Order,* 213.
244   "it is striking": Samuel Huntington, interview with Amina R. Chaudary in *Islamica Magazine,* Winter 2007, available at www.islamicamagazine.com/issue-17/an-interview-with-samuel-huntington.html.
244   "a central focus": Huntington, "The Clash of Civilizations?," 48.
244   "exploit differences and": Ibid., 49.
244   "The West will": Ibid., 49.
245   "To defeat a faith": Patrick Buchanan, "Coming Clash of Civilizations?," available at http://buchanan.org/blog/?p=400.
245   "Islam is a totalitarian": Robert Spencer, *The Myth of Islamic Tolerance: How Islamic Law Treats Non-Muslims* (Amherst, New York: Prometheus Books, 2004), 11.
246   "It is a purely": Maulana Syed Abul Ala Maududi, "Intellectual Subjugation: Why?" in S. A. A. Maududi, *West Versus Islam* (first written in 1934), 13–14.
247   "The leadership of": Sayyid Qutb, *Signposts on the Road,* 1964, from the introduction, 7–8.
247   "it is necessary": Ibid.
247   "If in the Muslim": Professor Khurshid Ahmed, "Islam and the New World Order," *The Muslim,* November 26, 1995.
248   "It is the height": "Secular Democracy Not Up to the Test," available at www.hizb.org.uk/hizb/resources/issues-explained/secular-democracy-not-up-to-the-test.html.
248   "But I could": Ayaan Hirsi Ali, *Infidel* (New York: Simon & Schuster, 2007), 272.
249   the most commented-on article: Huntington, *The Clash of Civilizations and the Remaking of World Order,* 13.
249   "For the past 2,000": Stephen M. Walt, "Building Up New Bogeymen," review of "Clash of Civilizations?" by Samuel P. Huntington, *Foreign Policy* 106 (1997): 183.
250   "just as cultural": Ibid., 184.
251   "In the Gulf war": Ibid., 185.
251   "can mobilize their": Ibid., 187.
251   "[T]his neglect of nationalism": Ibid.
251   "if we treat all": Ibid., 189.
252   "in this sense": Ibid., 189.
252   "From one of the": Fouad Ajami, "The Summoning," *Foreign Affairs* 72, no. 4 (1993): 2–3.
252   "Huntington buys Saddam": Ibid., 7.
253   "laid bare the interests": Ibid.
253   "their bids for power": Ibid.
253   "Civilizations and civilizational": Ibid., 9.
254   "development creates a": Robert L. Bartley, "The Case for Optimism: The West Should Believe in Itself," *Foreign Affairs* 72, no. 4 (1993): 17.
255   "the dominant flow": Ibid.
255   "militarized interstate disputes": Bruce M. Russet, John R. O'Neal, and Michae-

lene Cox, "Clash of Civilizations, or Realism and Liberalism Déjà Vu? Some Evidence," *Journal of Peace Research* 37 (2000): 591.

256 "also inhibits conflict": Ibid., 595.

256 "there is little evidence": Ibid., 596.

256 "not only fails": Ibid., 597.

256 "conflicts among split": Ibid., 600.

257 "Civilizations do not": Ibid., 602.

258 "states of different": Errol A. Henderson and Richard Tucker, "Clear and Present Strangers: The Clash of Civilizations and International Conflict," *International Studies Quarterly* 45 (2001): 328.

258 "mixed civilizations and": Ibid., 329.

258 "ignores the persistent": Ibid., 333.

259 "three perspectives: globally": Jonathan Fox, "Two Civilizations and Ethnic Conflict: Islam and the West," *Journal of Peace Research* 38 (2001): 459.

259 "overall there is little": Ibid., 463.

260 Despite Huntington's predictions: Ibid., 464–65.

260 "are a minority": Ibid., 466.

261 "This failure has": Huntington, *The Clash of Civilizations and the Remaking of World Order*, 114.

261 "comparative analysis of": Pippa Norris and Ronald Inglehart, "Islam & the West: Testing the Clash of Civilizations Thesis," John F. Kennedy School of Government, Harvard University, Faculty Research Working Paper Series, April 2002: 1.

262 "in recent years": Ibid., 11.

262 "contrary to Huntington's": Ibid.

262 "empirical findings directly": Ibid., 4.

263 "Similar to other": Russet, O'Neal, and Cox, "Clash of Civilizations," 589.

263 "If we treat all": Walt, "Building Up New Bogeymen," 189.

263 "Destructive conflict between": Richard E. Rubenstein and Jarle Crocker, "Challenging Huntington," *Foreign Policy* 96 (1994): 128.

264 "deep-rooted historical and cultural": Carl Gershman, "Should the United States Try to Promote Democracy in the Middle East?," speech given on August 11, 2007, available at www.ned.org/about/carl/carl081107.html.9.

266 "Member states of the": Mohammad Khatami, "Empathy and Compassion," speech given at U.N. General Assembly, September 8, 2000, available at www.iranian.com/Opinion/2000/September/Khatami/.

267 "Today, it is impossible": Ibid.

267 "cultural homelessness": Ibid.

267 "One goal of dialogue": Ibid.

268 "Dialogue is not": Ibid.

268 "In the past": "Executive Summary of the Publication of the Group of Eminent Persons Appointed by the United Nations Secretary-General on the Occasion of the United Nations Year of Dialogue Among Civilizations," available at www.un.org/Dialogue/summary.htm.

269 "The social and cultural": Peter Ford, "A Radical Idea: How Muslims Can Be European, Too," *Christian Science Monitor*, October 31, 2006.

269 "Sharia is a set": Quote from Ramadan in ibid.

269 "There is a clash": Ishtiaq Ahmed, "Debunking Civilizational Clash" (op-ed), *Daily Times* (Lahore), December 21, 2003.

6. RECONCILIATION

277 "The law revealed": Muhammad Iqbal, *Reconstruction of Religious Thought in Islam* (Oxford, UK: Oxford University Press, 1930), 139–70.

277 "The Muslims' aim": Fazlur Rehman, *Islam and Modernity* (Chicago: University of Chicago Press, 1982), 132.

277 "And here appears": Ibid., 134–35.

277 "It is the biographers": Ibid., 141–42.

278 **"The new step consists"**: Ibid., 145.
278 **"renewal has to start"**: Nurcholish Majdid, "The Necessity of Renewing Islamic Thought and Problem of the Integration of the Ummah," translated by Muhammad Kamal Hassan, in *Muslim Intellectual Responses to New Order Modernization in Indonesia* (Kuala Lumpur, Malaysia: Dewan Bahasa Dan Pustaka. Kementerian Pelajaran, 1980), 188–98.
278 **"Iman, Taqwa and the"**: Ibid., 217–33.
279 **"revealed religion is"**: Abdul Karim Soroush, "The Evolution and Devolution of Religious Knowledge," paper presented at Institute of Islamic Studies, McGill University, quoted in Charles Kurzman, *Liberal Islam* (New York: Oxford University Press, 1998), 250.
279 **"Text does not stand"**: Ibid., 245.
279 **"Revealed religion itself"**: Ibid., 246.
279 **"The whole history"**: Ibid., 248.
279 **"There is a need"**: Mohammed Arkkoun, "Rethinking Islam Today," Occasional Papers Series, Center for Contemporary Arab Studies, Georgetown University, Washington, D.C., 1987, 18.
279 **"Religious tradition is"**: Ibid., 20.
280 **"We must rethink"**: Ibid., 29.
280 **"All too many Muslims"**: K. H. Abdurrahman Wahid, "Right Islam vs. Wrong Islam," *Wall Street Journal,* December 30, 2005.
282 **"Once, when Prophet"**: Maulana Wahiduddin Khan, as reported in "Muslims Must Ignore Cartoons," *Pioneer* (India), March 2, 2006.
283 **"We may say that Islam's"**: Muhammad Khalid Masud, "Islam, Society, Modernity," colloquium paper, IRRI-KIIB, Brussels, Belgium, October 7–8, 2004.
283 **"Islam as a moral tradition"**: Ibid.
295 **But now almost**: Statistics from *Gulf News,* February 7, 2002.
296 **The Alaska Permanent Fund**: www.apfc.org/homeobjects/tabPermFund.cfm.
296 **"at least 25 percent"**: Alaska Constitution, Article IX, Section 15.
296 **This Alaskan fund**: www.apfc.org.
297 **"By no means shall"**: Quran 3:92.
297 **"And they ask you"**: Quran 2:219.
297 **"Who amasses wealth"**: Quran 112:274.
297 **"And keep up prayer"**: Quran 2:110.
298 **"Surely they who believe"**: Quran 2:277.
302 **"The world situation"**: George Marshall, speech given at Harvard University, June 5, 1947.
302 **"it is logical"**: Ibid.
311 **The Pew Research Center**: Andrew Kohut, "America's Image in the World: Findings from the Pew Global Attitudes Project," testimony before the Subcommittee on International Organizations, Human Rights and Oversight, Committee on Foreign Affairs, U.S. House of Representatives, March 14, 2007, available at www.pewglobal .org/commentary/display.php?AnalysisID=1019.
311 **"After Iraq, many"**: Ibid.
312 **"acknowledged that one"**: Edward P. Djerejian, "Changing Minds, Winning Peace: A New Strategic Direction for U.S. Public Diplomacy in the Arab and Muslim World," Report of the Advisory Group on Public Diplomacy for the Arab and Muslim World, October 1, 2003, available at www.state.gov/documents/organiza tion/24882.pdf.
312 **"the weapons of"**: James Traub, "Persuading Them," *New York Times,* November 25, 2007.